SCIENCE TEACHER EDUCATION

Science & Technology Education Library

VOLUME 10

SCOPE

The book series *Science & Technology Education Library* provides a publication forum for scholarship in science and technology education. It aims to publish innovative books which are at the forefront of the field. Monographs as well as collections of papers will be published.

The titles published in this series are listed at the end of this volume.

Science Teacher Education

An International Perspective

Edited by

SANDRA K. ABELL

Purdue University,
West Lafayette,
Indiana, U.S.A.

Published in cooperation with the Association for the Education of Teachers in Science

KLUWER ACADEMIC PUBLISHERS
DORDRECHT / BOSTON / LONDON

A C.I.P. Ctalogue record for this book is available from the Library of Congress.

ISBN 0-7923-6455-4

Published by Kluwer Academic Publishers,
P.O. Box 17, 3300 AA Dordrecht, The Netherlands.

Sold and distributed in North, Central and South America
by Kluwer Academic Publishers,
101 Philip Drive, Norwell, MA 02061, U.S.A.

In all other countries, sold and distributed
by Kluwer Academic Publishers,
P.O. Box 322, 3300 AH Dordrecht, The Netherlands.

Printed on acid-free paper

Printed in the Netherlands.

Table of Contents

ACKNOWLEDGEMENTS

My guess is that, if I would have known then what I know now about what it takes to edit a book such as this, I might not have sat in Paul Kuerbis's little house on the Colorado College campus several years ago, so excited to start this project. It was Paul's insight, as then President of the Association for the Education of Teachers in Science (AETS), to explore issues of science teacher education worldwide. He asked me to edit an AETS sponsored book on the subject. As you can see, I agreed.

It has been my pleasure to work with authors from all over the world, many of whom I have met only through email conversations. It has been my privilege to learn about issues of science teacher education in their countries, and think about how their views influence my own. And it has been my honor to represent the AETS mission through editing this book.

I would be remiss if I failed to mention the support I have received from Purdue University to carry out this project. Purdue funded two different undergraduate education students who worked on this project. First Amy Jacks and later Pat Van Dermay learned more than they expected about the ins and outs of copy editing and indexing. Moreover, they provided undying enthusiasm for the project and for their own development as teachers. I have enjoyed watching this development up close. A special thanks goes to Karen Hearn, who figured out how to transform manuscripts into camera-ready copy, and put all of the pieces together. The beautiful book in front of you was crafted at her keyboard. Thanks.

S.A., West Lafayette, Indiana, USA

LIST OF REVIEWERS

The AETS members listed below served as reviewers for proposed chapters to *Science Teacher Education: An International Perspective*. I would like to thank them for their conscientious reviews and for their committed service to AETS. Thanks also to the two anonymous reviewers who examined the entire manuscript for Kluwer; your suggestions have led to significant improvements throughout the book.

S.A.

Dale Banks, Saint Mary's College, USA
Michael Cohen, Indiana University - Purdue University at Indianapolis, USA
Angelo Collins, Vanderbilt University, USA
Ed Donovan, University of South Carolina—Spartenburg, USA
Judy Egelston-Dodd, Rochester Institute of Technology, USA
Alejandro Gallard, Florida State University, USA
Mark Guy, University of North Dakota, USA
R. Lynn Jones, University of Texas – Austin, USA
David Kumar, Florida Atlantic University, USA
Chin-Tang Liu, South West Missouri State University, USA
Ann Haley MacKenzie, Miami University of Ohio, USA
Bruce Munson, University of Minnesota – Duluth, USA
Sherry Nichols, East Carolina University, USA
Sharon Parsons, San Jose State University, USA
J. Preston Prather, University of Tennessee – Martin, USA
Steve Rakow, University of Houston - Clear Lake, USA
Laura Rogers, University of North Carolina – Wilmington, USA
James Russett, Purdue University – Calumet, USA
Art St. George, National Science Foundation, USA
Peggy Tilgner, Wartburg College, USA
Meta Van Sickle, University and College of Charleston, USA
Lawrence Wakeford, Brown University, USA
Amanda Woods McConney, Western Oregon State College, USA
Carla Zembal-Saul, The Pennsylvania State University, USA

SECTION I

INTRODUCTION

Chapter 1

International Perspectives on Science Teacher Education

An Introduction

Sandra K. Abell
Purdue University, USA

Science education has seen its share of reform efforts come and go. For the most part, these reforms have centered around changes in science curriculum and instruction. We are now in the midst of yet another wave of reform, this one sparked by the publication of science standards in several countries. This time, however, the reform efforts are attending to a feature of reform that has often been forgotten: science teacher education. Reformers have realized that new curriculum or innovative instructional techniques need teachers to carry them out. Thus focusing on reform in science teacher education will be crucial to the success of other science education reforms.

The authors of the pieces in this volume have witnessed their share of reform efforts in their countries, some driven by government policies, some by institutional initiatives, and others by the researchers themselves. Collectively the writers paint a promising picture of science education internationally. Their picture is one of change and progress, of commitment and hope. Their stories are presented in the spirit of capturing history and moving forward to inform the future.

OVERVIEW

The opening section of this volume, "Policy and Practice in International Science Teacher Education," takes us to four different nations to examine the development of teacher education and the directions for the future. Appleton, Ginns, and Watters discuss elementary science teacher education

S.K. Abell (ed.), Science Teacher Education, 3–6.

in Australia, which has moved from a one year program to a four or five year qualification. Borghi, De Ambrosis, and Mascheretti illustrate the reform in science teacher education in Italy with the case of physics teacher education. They challenge teacher preparation programs to provide a wide range of coursework in education, science, and the history and nature of science to engage students in the most current views of best practice. BouJaoude details science teacher education in a variety of higher education institutions in Lebanon and provides a set of recommendations for the future based on the strengths and weaknesses of these programs. Iqbal and Mahmood summarize the science teacher education situation in Pakistan, providing recommendations for higher education to work with teacher inservice agencies to educate future science teachers. The issues that these authors discuss, although specific to their own national contexts, are by no means unique. Their solutions have the promise of informing other nations in the midst of science teacher education reform.

The next section, "Making Sense of Science Teacher Learning," highlights research on student learning in teacher preparation programs. Weinberger and Zohar discuss the use of a curriculum designed to improve higher order thinking skills in a junior high teacher preparation program in Israel. Their study demonstrates that preservice teachers can learn how to develop higher order thinking in their students as a result of developing their own thinking skills. Baird, Brodie, Bevins, and Christol examine the student teaching experience of secondary science teachers in the United States and the United Kingdom. Their work has led to the creation of a model for the student teaching experience. Abell and Jacks tell the story of one student, early in her elementary teacher education program, and how she learned to think like a teacher while participating in a Study Abroad internship in Honduras. These authors are generating knowledge about science teacher education that can inform reform efforts globally.

The final section, "Cross-cultural Perspectives on Science Teacher Education," reports on several international partnerships that are generating new knowledge about science teacher education. Koch and Calabrese Barton tell the stories of Egyptian teachers engaged in a US teacher enhancement project. Their stories help us understand the cross-cultural issues that come into play when instructors and administrators work with teachers from different countries. Pedersen, Bonsttetter, Rioseco, Briceno-Valero, O'Callaghan, and Garcia discuss the political and cultural climate for change in Chile, Bolivia, and Venezuela, and how partnerships with higher education institutions in the US have contributed to science teacher education reform in these countries. The book ends with a chapter by Tippins, Nichols, and Bryan, in which they present the voices of their international colleagues—Amadou, Chun, Ikeda, McKinley, Parker, and

Herrera-- in an attempt to understand scientific literacy and science teacher education from a global perspective.

THEMES

What I have learned from these stories is that the problems faced in science education in a given country, although unique to its history, politics, and culture, also share commonalties with other places. Furthermore, the solutions to these problems, both envisioned and enacted, reveal common themes in our science teacher education work. I would like to explore some of these themes briefly, before setting you off on your own journey through these stories.

Not surprisingly, no place has "arrived" at the pinnacle of science education. The authors, whether from developed or developing countries, describe situations where science education and science teacher preparation are less than adequate for the science literacy goals embraced by governments, citizens, and educators. In many countries, a portion of school-aged children receive little or no science education—because they do not attend schools; because the schools they do attend are lacking in resources, including prepared teachers; or because their language and cultural background exclude them from school science. In many places worldwide, science educators and government officials have recognized the weaknesses in their systems and are making strides to ameliorate the problems.

One science education reform strategy that is clearly at the top of the list, again in both developed and developing countries, is to restructure science teacher education. The authors in this volume agree that it takes time to learn to be a teacher. They agree that teachers must understand science concepts, principles, and the nature of science; how students learn; and science curriculum, instruction, and assessment. They agree that teachers need time in the field, working with students and reflecting on practice. They recognize that teachers are the key to reform and that improved teacher education is an essential feature of successful reform. The other day I heard a radio commentator suggest that, to improve schools in the US, we should close all faculties of education, stop preparing teachers per se, and ask students with bachelor's degrees in any field to teach our children. As a teacher educator, I was appalled by the simplemindedness of this proposal and angered by the lack of understanding of teacher education and the profession of teaching. Many of the initiatives described in this book have the potential to make a real difference in science education around the world.

I challenge all of us who care about schools to create and enact policies and practices aimed at challenging business as usual and changing the status quo.

We should not expect these changes to take place quickly. The success of the reforms will depend not only on the quality of the education program offered, but also on the political, economic, and cultural climates in which they take place. The American Association for the Advancement of Science (1993) was farsighted to recognize that deep and lasting change takes time, and that the year 2061 is a realistic timeline to achieve the science education reforms initiated in the latter part of the 20th century in the US. In other places, however, reform measures may last only as long as the current political regime. What this means for science teacher educators is that we must be forever vigilant, looking for the best opportunities to take action and have an effect. And although we must think globally about the issues and values in science education, we must act locally to affect our particular contexts.

This book is an attempt to highlight some themes in science education reform and describe various efforts to improve science teacher education around the globe. I advise the reader to search for your own connections to the policies, practices, investigations, and stories presented here, and then use them to take action on your own situation. We can understand our own situations better by understanding the situations of others. Collectively we can then continue to change the global landscape of science teacher education.

REFERENCES

American Association for the Advancement of Science. (1993). *Project 2061: Benchmarks for science literacy.* New York: Oxford University Press.

SECTION II

POLICY AND PRACTICE IN INTERNATIONAL SCIENCE TEACHER EDUCATION

Chapter 2

The Development of Preservice Elementary Science Teacher Education in Australia

Ken Appleton[1], Ian S. Ginns[2], and James J. Watters[2]
[1]Central Queensland University, Australia: [2]Queensland University of Technology, Australia

Abstract: This paper describes the emergence of the contemporary structure of preservice primary and elementary science education in Australia. We present an historical account of the development of current programs and an analysis that reveals the major trends and influences that have molded the current situation. Major changes have occurred since the late 1970s but the last decade has seen revolutionary restructuring. We discuss the relevant literature, drawing on research reports, reports of national and state governments and other reviews. We analyze how these trends and influences have shaped education policy and preservice programs in universities. We explore as well, emerging trends and implications for future developments.

Teacher preparation in primary and elementary science in Australia has undergone a major transformation over the last thirty years, particularly during the last decade. Like all education, teacher education is influenced by political processes, both within the profession and within the community. In this chapter, we trace the changes that have occurred, the political and educational contexts that have framed these changes, and the effects they may have had on the teaching of science. Most notable has been the increasing intrusion of community politics into teacher preparation, moving it from an activity largely controlled by the teaching profession to the political arena.

The chapter assumes the form of an historical narrative, recounting key events both internal and external to the profession which have shaped elementary science teacher preparation, and a critical analysis of the changes that have resulted. The narrative is told from the perspective of the authors, who have been involved in the field over the period discussed. Since our main area of interest lies in the preparation of teachers for primary and elementary teaching, we have concentrated our story on this aspect of

S.K. Abell (ed.), Science Teacher Education, 9–29.

science teacher preparation. Our story may also be colored by the fact that we live in the state of Queensland, so our account may not reflect fully events in other Australian states and territories.

THE CONTEMPORARY SITUATION

Currently, preservice elementary teacher education in Australia is undertaken mostly in Faculties of Education in the majority of the 39 universities. While there are some subtle differences in the broad structure of programs, the commonalties outweigh the differences. Most teachers in early childhood settings and elementary schools are accredited through a four year Bachelor of Education degree program. Alternatives for students who have completed a bachelor's degree in a discipline and subsequently decide to follow a teaching career include a one year Graduate Diploma of Teaching, a graduate two year bachelor's program, or a two year master's program. Therefore, the majority of commencing teachers will have undertaken at least four years of university study, often wholly within a Faculty of Education. By comparison the standard entry into the profession in 1960 was completion of a one year certificate of teaching awarded by a teachers college.

Before we explore the background to the development of teacher education fully, we provide for the reader a brief overview of the demographic and political contexts of education in Australia up to and including this point in time.

The Demographic and Political Context

Australia is an island continent with over 18 million inhabitants. Although Australia occupies a large area, much is desert to semidesert, resulting in concentrations of population around the more fertile southeast, along the east coast, and in the southwest corner of the continent. The remainder is sparsely populated. Predominant industries are agriculture, mining, and tourism.

The system of government in Australia evolved from the establishment of colonies in different geographic locations. The first was in Sydney in 1788, and over the ensuing decades other colonies began either as British penal settlements, or as centers for settlers who spread rapidly in search of new lands. The main settlements eventually became centers of government for separate states: New South Wales, Victoria, Tasmania, South Australia, Western Australia, and Queensland. Each state provided basic services such as public education, though churches had originally assumed the sole responsibility for this. In 1901, the Commonwealth of Australia was established as a federation of states. The power and influence of the federal government have increased progressively in terms of its influence on national policy, particularly through taxation and distribution of revenue.

Education in Australia

The provision of free compulsory education was an early initiative of the colonies and continues to be the responsibility of each state and territory (a territory is an area not originally party to the formation of the Commonwealth but separate from the states). Currently, the Commonwealth government provides supplemental funds to the states and territories for the elementary and secondary sectors, and assumes funding responsibility for the tertiary sector. Free tertiary education was introduced in 1974 but partial tuition fees were reintroduced in 1989.

Elementary education, or primary school as it is called in Australia, covers Years 1 to 6 (ages 5 to 11) in most states and territories, and Years 1 to 7 in two others, including Queensland. Compulsory schooling extends to 15 years of age, usually to the end of Year 10, with upper secondary covering Years 11 and 12. On completion of high school students may achieve some form of tertiary entrance score which Universities use to determine entrance eligibility. Primary education, or early childhood education (ages 3 to 8) as it is called in Australia, also varies from state to state. Child care and kindergarten are often available for children under 4 years of age, and most states and territories have preschool available for children from 4 years of age.

Administration and the providing of resources for primary and elementary education is influenced by differences in population distribution, political ideology, and physical distance so there is considerable variation in educational structure and curriculum among the states and territories. However, within each state, curriculum development is a responsibility of a central body and while individual schools have some flexibility, programs and structures are relatively uniform.

Our historical account of developments in elementary science teacher preparation in Australia follows.

THE SITUATION PRIOR TO THE 1980s

Approximately thirty years ago the main political processes influencing science teacher preparation in Australia were those emanating from within the profession, such as teachers, teacher educators, and the elementary science curriculum itself.

Elementary Science Education

Before the 1960s, the only science component of the elementary school curriculum was nature study. About this time, each state began to introduce a more general science syllabus. However, these programs were not well resourced, teachers were not aware of how to teach science, and as a relative newcomer to the curriculum, science teaching received a low priority. The states had their own science syllabuses with consequential differences in emphasis. Elementary science syllabuses produced during the period 1960 to 1990 tended to be based on curriculum development ideas from other countries (e.g., SAPA, SCIS, ESS from the US and Science 5/13 from the UK), and often did not take account of contemporary Australian developments and research findings. In the 1970s, concerns began to be expressed about the quality of elementary science teaching (Symington, 1974; Varley, 1975).

Elementary Teacher Preparation

Forty years ago, the main route to being an elementary teacher in Australia was to receive several years of apprenticeship under the supervision of an experienced teacher after completing Year 10. However, as demand for more and better trained teachers increased, governments began to introduce scholarships as an incentive to boost enrollments. Training was conducted in teachers colleges, where the typical program was one year in length following completion of Year 12. This later increased to two years in the 1960s, and scholarship holders were bonded to work in government schools for several years after graduation. An effect of scholarships being available was to increase the number of men entering the profession to 40 or 50 percent.

In 1971, as a result of a review of higher education (Martin, 1964), the Commonwealth government negotiated a financial arrangement with the states to fund tertiary education. A binary tertiary system was established, consisting of the established universities, and a new group of professional colleges called Colleges of Advanced Education (CAEs). The teachers colleges and industry-training colleges, such as Institutes of Technology, were transformed into the new CAEs, and were encouraged to broaden their program offerings into other professions and fields of study such as business. At this time, elementary teacher preparation became a 3-year diploma program following Year 12. Inservice upgrading qualifications also became available for practicing teachers to obtain a diploma or degree. Some CAEs offered a preservice 4-year degree, but this was the exception. Programs in CAEs were nationally accredited through state committees, which maintained monitoring and advisory roles. Other state committees had a role in registering teachers in those states where teacher registration legislation was introduced. Teachers were well represented on the accrediting and registering bodies and most education staff at CAEs were drawn from the teaching force, so the shape and overall content of teacher education programs were largely in the control of the teaching profession. Thus teacher education at that time was firmly embedded in local issues and modeled on apprenticeship training.

Research in Elementary Science

There was little interest in research into elementary science education or elementary science teacher education until academic staff specializing in elementary science were appointed to CAEs. However, the amount of research conducted was limited, since CAE staff were expected to be engaged in teaching and were not encouraged or funded to undertake research. Research was therefore seen as the prerogative of the universities, though none were then engaged in elementary teacher preparation. Despite this, a small number of CAE staff regularly attended annual meetings of the Australian Science Education Research Association[1] (ASERA), which became an important professional link for them, and as such was to later exert considerable influence on elementary science teacher preparation. Several research reports by these people focused on elementary teachers' practices in science teaching (e.g., Appleton, 1977; Henry, 1977; Skamp & Power, 1981; Symington, 1974; Varley, 1975), painting a fairly dismal picture of large scale avoidance of science teaching, and teaching dominated by teacher lectures, television, and whole class discussion.

THE SITUATION IN THE 1980s

Since 1980, a number of related influences have had considerable impact on elementary science teacher preparation. In many cases these influences can be associated with specific events.

[1] The Australian Science Education Research Association (ASERA) was formed in 1970, and renamed the Australasian Science Education Research Association in 1990.

The Research Interests of Academics

The emergence of a small group of College of Advanced Education academics who conducted research into elementary science and elementary science teacher education had a significant effect on curriculum development in some teacher preparation programs. Equally important was the continuing growth of ASERA and its related publication, *Research in Science Education*, as a forum for reporting research and discussing its implications. The association held a philosophy of supporting emergent researchers and conducting low cost conferences with considerable opportunity for discussion. This was important for researchers in elementary science as they had all moved into academia following a teaching career and had subsequently undertaken research training, often part time, and initially at the master's level.

As mentioned above, early research tended to focus on the teaching of science, but there was a subsequent focus on preservice teacher education students as CAE academics began to research their own teaching and teaching contexts (e.g., Appleton, 1983, 1984; Dooley & Lucas, 1981; Garnett & Tobin, 1984; Ginns & Foster, 1983; Hope & Townsend, 1983; Skamp, 1987; Symington, 1980, 1982). This research consistently showed that preservice teachers lacked confidence in teaching science, and that specific interventions in science content or science methods courses had positive effects. A result of this research activity was that science teacher preparation programs came under close, reflective scrutiny in several key institutions, so that they were open to new initiatives and ideas that emerged during the next decade. The foremost of these was constructivism.

Constructivism

In the early 1980s, a series of events occurred that would have a significant effect on elementary science teacher preparation. The science classroom research projects—the *Learning in Science Project* and later the *Learning in Science Project (Primary)* began in New Zealand. These projects focused on children's learning in science at lower secondary and elementary grades respectively. Progress reports and findings were presented at the annual conferences of the ASERA, and were published in *Research in Science Education* by the project leader, Roger Osborne and his research teams. Key international visitors such as Rosalind Driver, John Gilbert, and Merl Wittrock also became involved. These developments had two effects. Firstly, reports of elementary science education research received a much higher profile in the Australian science education research community, and some of the research-oriented CAE science educators became involved in the projects as visiting academics, co-researchers, and consultants. Secondly, since the projects took a constructivist view of learning (Biddulph & Osborne, 1984; Osborne & Freyberg, 1985), this perspective was slowly introduced into the content and practice of tertiary science teacher education. For instance, one of the authors (KA) spent a 6-month sabbatical during 1984 working with the *Learning in Science Project (Primary)* team, and on returning to his institution began to explore how to change his science education (methods) courses to incorporate teaching and learning using constructivist ideas.

The involvement of Australian science educators generated more Australian based research on elementary science and elementary science teacher education over the next few years, which in turn influenced teacher preparation programs at other CAEs. A growing interest in constructivism among the international science education community resulted in a snowballing effect over the remainder of the decade that saw constructivist views of learning transform most of the science teacher education programs in Australia. However, paralleling these advances in educational research, theory, and practice were a number of political decisions during the 1980s which markedly influenced teacher education. These decisions culminated in 1989 with major policy initiatives that can be seen as a watershed in Australian science teacher education.

Government Policy

In the early 1980s, forced amalgamations of some existing CAEs by the Commonwealth government resulted in new CAEs with large multicampus operations. Although having no immediate effect on teacher education, except for reducing the number of programs offered, the move signaled an increasing interventionist role of the Commonwealth government in education. At the state government level, a decision was made to cease providing scholarships for students enrolled in teacher education programs, consequently decreasing the percentages of men entering the profession to as low as 10 to 20 percent. This action, combined with a shrinking job market, saw student numbers fall dramatically. The consequential smaller classes, however, made it easier for new ideas about teaching using constructivist ideas to be trialed.

In the mid 1980s, the Commonwealth government, driven by the ideology of the Labor Party which was in power at that time, introduced new policy initiatives such as social justice and equity. These were deemed as priority issues that attracted funding support for inclusion in teacher education programs. Indirect political control was exerted over schools as well, by making some supplemental funding conditional. National economic growth also assumed importance, a key factor of which was the restructuring of the economy to enhance Australian competitiveness, especially in manufacturing industries that depended on science and technology. However, political dissatisfaction with the school system increased with mounting criticism of literacy and numeracy outcomes, continuing reports of poor elementary science practices (e.g., Bennett, 1984; Owen, Johnson, and Welsh, 1985), and declining science enrollments in senior high school. A consequence was a Commonwealth government commissioned review of science and mathematics teacher preparation (Department of Education Employment and Training [DEET], 1989). The ensuing report noted a considerable variety in science teacher education courses, but was highly critical of some institutions that had little compulsory science education in their programs, and those which persisted with outdated views of learning and teaching. Subsequently the report was used effectively in many institutions as a lever to increase time and resource allocation to elementary science teacher education. It also spawned increased research activity into elementary science and science teacher education. Much of this focused on constructivism and the classroom application of constructivist principles (e.g., Appleton, 1993; Goodrum, 1993). Research into elementary science teacher education continued to show that student teachers lacked confidence to teach science (Watters & Ginns, 1994), were well disposed to science as part of the curriculum (Farnsworth & Jeans, 1994), and could be positively influenced by appropriate experiences in science or science methods courses (Appleton, 1991; Kirkwood, Bearlin, and Hardy, 1989; Ginns, Watters, Tulip, and Lucas, 1995; Hand & Peterson, 1995)

Another outcome of the DEET (1989) report was that substantial funding for professional development of primary and elementary teachers in science was made available. Two notable programs were the *Primary and Early Childhood Science Teacher Education Project* (PECSTEP) (Hardy, Bearlin, and Kirkwood, 1990; Kirkwood, Bearlin, and Hardy, 1989) and the *Sci-Tech Inservice Project* (Napper & Crawford, 1990). Both programs placed a heavy emphasis on constructivist ideas. PECSTEP also took into account gender issues, which were recognized as an essential component of elementary science teacher education, given the high percentage of women in the workforce. While these programs had a predominantly inservice focus, they had immediate effects on shaping preservice programs in the project institutions and others (e.g., Appleton, 1991, 1995).

Further, the Australian Education Council (AEC), a body consisting of all state, territory and commonwealth ministers of education, forged an agreement (referred to as the Hobart declaration) in 1989 to work toward a common curriculum framework for the nation in eight Key Learning Areas.

These included both Science and Technology as two separate areas. Almost all of the participating politicians were of the same political persuasion, which facilitated achievement of consensus about the goals of education. There was also strong agreement that the curriculum for each Key Learning Area should be outcomes based, reflecting international trends. This first concerted move toward a centralized curriculum in Australian history led to the commencement of work on curriculum frameworks in science and technology that could be used by each state and territory as the basis for common curriculum development.

A decision to restructure the tertiary sector was again reached by the Commonwealth government near the end of the decade (Dawkins, 1987). As the distinction between universities and CAEs had become increasingly blurred, a unified national university system was proposed. Under the new unified system, many institutions amalgamated, and CAEs disappeared as they were either absorbed into existing universities or were restructured as universities.

A number of important effects were generated by the government policy initiatives, which flowed into the 1990s:

a) All elementary teacher preparation now occurred in a university environment. The three year diploma programs were replaced by degree awards, thus enhancing the status of the profession.
b) The monitoring and accreditation function of state committees also disappeared or diminished as the new universities and newly transformed education faculties in existing universities assumed self-regulation. This, together with the moves toward offering degrees and the review of science teacher education (DEET, 1989), resulted in an enhanced offering of science in many elementary teacher education programs. New political moves in some states also saw the removal of state teacher-registration bodies, thus allowing some universities to reduce practicum (field work in schools) requirements. This was done to save money in a framework where teachers were paid to supervise student teachers during field work.
c) Academic staff from the now defunct CAE sector found themselves having to adjust to a new university environment. Many retired or alternatively decided to upgrade their qualifications, as they were expected to engage in research. The number of researchers inquiring into elementary science or elementary science teacher education increased dramatically, and there was a shift to greater internationalization of research output. Academic staff, now becoming better informed by recent international research, were more proactive in pursuing their own research interests in the Australian context. They were encouraged to

report their findings at national forums like ASERA and at international meetings such as those of the National Association for Research in Science Teaching or the American Educational Research Association. Hence, elementary science teacher preparation had now become influenced by academics with strong theoretical backgrounds that were embedded in a contemporary national and international research culture. This contrasted with the strong professional influence that had inspired earlier science teacher preparation.

At the beginning of the 1990s, the stage was set for dramatic changes in elementary science teacher preparation. But the government was not finished yet!

THE SITUATION IN THE 1990s

The 1990s have been a period of continuing political intervention. There has been an increased emphasis by Commonwealth and State governments on having their political agendas reflected in the various education systems. Issues such as national competitiveness, economic rationalism and social equity continue to dominate political thinking and decision making. These political objectives have influenced the development of science curriculum in schools and the structure of the tertiary education sector, both of which have influenced preservice science teacher education.

Reshaping of the Curriculum and Impact on the Profession

Following the Hobart Declaration, a national collaborative effort under the guidance of the AEC focused on producing statements and profiles (outcome statements) in each of the eight agreed Key Learning Areas in order to alleviate perceived problems concerning imbalance in curriculum structures across states and territories. The science statement and profiles were designed to provide a common framework for curriculum development and provide guidelines for reporting on students' progress in science. However, during the early 1990s, the balance in the political climate in Australia changed dramatically and the ideological consensus of the AEC was diminished. By July 1993, attempts at implementing a uniform national curriculum were disbanded and each state and territory began its own adaptation of the Australian Science Statement (Curriculum Corporation, 1994a) and Profiles (Curriculum Corporation, 1994b).

While Tasmania, South Australia, Western Australia, the Northern Territory, and the Australian Capital Territory have incorporated broadly the statement and profiles into their own programs, the Queensland state sector has only just developed a syllabus that is framed around the science statement. Victoria wrote their own completely restructured version of the science statement and profiles, and New South Wales has been grappling with how to accommodate the separate Key Learning Areas of Science and Technology with a very recent combined elementary Science and Technology syllabus.

The Australian Science Statement has influenced commercial publications, which have adopted the profile or statement as an organizing framework. However, the absence of a unified approach to the implementation of the Australian Science Statement and Profiles has remained a major impediment to full engagement by teachers in the initiative. Nevertheless, the documents have had an important impact on the teaching of science and on science teacher education in Australia.

Apart from the flow-on effects of the statement and profiles for science teacher curriculum change, two other developments had implications for science teacher preparation:

a) Commonwealth funding was made available to teachers throughout Australia to familiarize themselves with the national statements, irrespective of the direction their own state education sectors were following. In particular, the funding was targeted at teacher subject associations such as the Australian Science Teachers Association (ASTA). This strategy was a deliberate attempt to circumvent state government appropriation of funds, and to ensure the funds were spent according to Commonwealth government priorities. ASTA consequently received over a million dollars to develop a professional development program around the Australian Science Statement and Profiles. For example, the *South Australian Primary Science Project* involved extended workshop courses on selected areas of the Australian Science Statement and Profiles as part of the South Australian Science Teachers Association contribution to the national professional development initiative (Nikkerud & DUnienville, 1995). Similar schemes were implemented in other states and in some instances brought together in close collaboration tertiary educators and professional teacher associations.

b) National science and technology conferences for elementary school educators commenced in 1993 on a biennial basis. Keynote addresses and workshops at these conferences have focused on the Australian Science Statement and Profiles and provided opportunity for debate about the issues associated with the implementation of outcomes based science and technology education in elementary schools.

The Australian Science Statement and Profiles have had a profound effect on elementary science teacher education. Firstly, the identification of science as a Key Learning Area, and the curriculum development that subsequently emerged, provided an added impetus for the inclusion of more time on science education in preservice programs, although, as we shall see, institutional and other factors were at work to counteract this. Secondly, the professional development programs that sprang up around the country had spinoffs into preservice programs. In many instances, university academics teamed up with the subject associations, so they were able to use the wisdom obtained from the professional development programs in their preservice teaching. Thirdly, many elementary teachers and elementary science educators were able to discuss issues as never before in the forums provided by the science and technology conferences. This too provided a means for the review of science teacher education programs.

An Independent Curriculum Initiative

Work on an independent curriculum initiative was commenced by a university research team headed by Denis Goodrum. In 1992, they reported on a research project where teachers had been given recently published (overseas) curriculum materials during focused professional development programs (Goodrum, Cousins, and Kinnear, 1992). The teachers subsequently trialed the materials in their classrooms, with considerable ongoing support. Using this successful trial, Goodrum succeeded in obtaining the support of the Australian Academy of Science (AAS) for a curriculum development project. Government and corporate sponsorship resulted in the publication of *Primary Investigations* (AAS, 1994). Funding for a professional development program for the curriculum and for training an extensive network of trainers was also obtained. This was the first major elementary science program available nationally, and was subsequently adopted widely by Australian elementary schools in some states, with the greatest uptake closest to the key trial centers. Although developed independently of the Australian Science Statement, supplementary documentation was provided that articulated links between its organization and structure and the Profiles. *Primary Investigations* and elements of the professional development program have found their ways into many teacher preparation programs, just as the curriculum developments of the 1960s, such as SAPA and SCIS, were popularized in the United States.

Attempts at developing a national curriculum and providing national professional development orchestrated by political initiatives have been only marginally successful. Top down approaches have been seen by the professionals in a cynical light. As indicated above, the Australian Science

Statement and Profiles have impacted science teacher preparation, but this has been problematic as each state has interpreted or adopted them in different ways. Although there is some consistency in the overall structure of preservice programs, science education courses have, as a consequence of government policy initiatives and reports, remained fragmented with no consistent approach or structure being adopted by universities, apart from an emphasis on constructivism. While advances have been made, the proactive roles of government and industry groups in defining policy for the tertiary sector and its structure and operation appear to have been detrimental to the development of a consistent national strategy for science teacher preparation. This matter is expanded upon in the next section.

Impact of Reshaping the Tertiary Sector on Teacher Education Programs

As noted previously, the momentum of Commonwealth initiatives in higher education increased greatly in the late 1980s and reform of the higher education sector was a major item on the agenda of the Commonwealth Labor government for continuation into the 1990s. One of the recommendations of the review of science and mathematics teacher preparation (DEET, 1989) was that the Department of Education Employment and Training should commission a study to report on the implementation of all the recommendations of that review. Findings presented in the report of this study indicated that a number of universities had increased the time for science content and curriculum courses but the increases in many cases were illusory in the sense that although there were now two courses of study instead of one, the increase in overall time devoted to science was minimal over an entire 4-year preservice program (Whitehead, Symington, Mackay, and Vincent, 1993). Symington and Mackay (1993) also concluded that there had been relatively little critical discussion among science educators of the assumptions underpinning the review. A clear response was increased attention to discipline (e.g., science) studies for preservice elementary school teachers with science content courses being developed by universities and in some places taught outside the education faculty. The main concern, also related to restructuring of universities and relocation of courses, was the extent to which it was appropriate for science education and science content courses to be offered in the same faculty, namely education. The case on one side argued for rational economics, while those charged with the preparation of teachers argued that a dual responsibility for curriculum and content would be more effective because teacher educators had greater interest in the whole professional development of elementary school teachers. In some universities there was a move to abandon science methods courses totally in favor of science content, though these moves may have been motivated more by institutional desires to reduce the size of large Education Faculties resulting from amalgamations.

Economic Rationalism

An increasing concern of the Commonwealth government toward the end of the 1980s and into the 1990s has been the escalating cost of the tertiary sector, mainly caused by high unemployment, high retention rates into Year 12, and consequential rapid increases in university enrollments. The first action by the Commonwealth was to reintroduce partial fees in 1989. These have been progressively increased in subsequent years. Other actions included lowering funding levels for universities, and more recently requiring universities to find supplemental sources of funding. The combined effects of greater enrollments with less funding has had a considerable impact on elementary science teacher courses. Staff-student ratios have increased, with academics being required to take more students overall and larger classes. Consequently, it has become difficult to sustain courses with heavy emphases on laboratory work, discussion, personal encouragement, small group work, inquiry and investigatory work, and assignment/project work.

THE LATEST VIEW

The outcomes based trend in curriculum development arising from the Hobart Declaration, reinforced by other Commonwealth government sponsored inquiries into education and training such as the Finn (1991), Carmichael (1992) and Mayer (1992) reports, was reflected in Commonwealth government thinking about teaching and teacher preparation. *The Key Competencies Report* (Mayer, 1992), foreshadowed by the Finn and Carmichael reports, was based on the premise that the role of general education was for the main purpose of employment. The Mayer report was notable because of its political agenda of transmission of government preferences and values for defining a sustainable future for a multicultural Australia through educational institutions.

Arising from the Mayer report, a research and development project was launched to investigate ways to improve the quality of teaching and learning in Australian schools. Funding for the project was provided by school authorities, teacher unions, the Commonwealth government, and the

Australian Council of Trade Unions. One outcome of the project was the publication of a National Competency Framework for Beginning Teaching (National Working Party on Teacher Competency Standards, 1996) which described the qualities, knowledge, attributes and skills needed by teachers. The Framework was distinguished by its national nature, its research base, and its teacher-focused development process. Universities, for the most part, have placed the implementation of the Framework in their strategic plans, but there has been little overt integration of the framework into teacher education programs to this date.

Elementary science education continues to be seen as an important means of increasing the level of the Australian community's scientific literacy and empowering individuals in a technological society (Australian Science Technology and Engineering Council [ASTEC], 1997; Fensham, 1994). The ASTEC report has argued that as the impact of science (and technology) on our daily lives continues to grow, so too must the level of our society's scientific and technological literacy. Indeed, being literate and numerate in the purest sense of the terms, will surely not be sufficient to live as informed citizens in any western society in the 21st Century (ASTEC, 1997).

The main findings of the report were that much has been achieved in elementary science education over the past ten years, but more remains to be done. Elementary science enjoys a strong level of support among school principals, parents, teachers and children. However, many primary teachers (both recent graduates and experienced) are not totally confident about teaching science, and typically, only 45 to 60 minutes (or about 4% of the teaching time) are allocated in the weekly timetable for science. The report argued that action is required to address the low teacher confidence in teaching elementary science and suggested that Faculties of Education have paid insufficient attention to the science content of preservice programs. It is significant that this study, and other recent research (e.g., Appleton & Kindt, 1997; Skamp, 1991), found that the teaching of elementary science remains problematic, despite considerable change in elementary science teacher preparation over the previous decade. In the report, the assumption was made that low self-confidence can be addressed just by increasing science content, a conclusion which we feel has yet to be justified (Appleton, 1995; Smith, 1997). While the report had again singled out university teacher education programs for criticism, we have begun to explore whether there are other factors exerting greater influence on elementary teachers' science teaching behaviors (Appleton & Kindt, 1997; Ginns & Watters, 1997).

The future impact of the ASTEC Report on elementary science teacher preparation is difficult to predict, since the committee was moved to a different government portfolio shortly after presenting its report. In its new location, we suspect it will be less likely to influence government education policy.

While government bodies such as ASTEC have decried the status of science education, economic and government ideological positions constrain the capacity of universities to react. The application of innovative and theoretically sound educational practices in preservice programs requires a number of initiatives that are expensive. However, increased large group lecturing, devolution of responsibility for face-to-face teaching to part time or casual staff, reduction in practice teaching opportunities for students, diminished resources for preservice programs, and the increasing administrative and research demands on academic staff are placing considerable pressure on maintenance of quality in teaching. In the last two years, the tertiary sector has been severely affected by savage Commonwealth funding cuts which have not been compensated for by an increase in private sector funding, which is difficult to obtain in the domain of teacher education.

IMPLICATIONS

A number of achievements have been realized for preservice primary and elementary science education in Australia during the past thirty years. Previously, teacher education was closely linked to the teaching profession and was oblivious to many issues and trends emanating from around the world. Today, preservice programs are strongly influenced by contemporary international research strengthening the nexus between research and practice. Indeed, we have seen the change from teacher training to teacher education, with connections to the profession remaining strong but now manifested through liaisons with the professional science teacher associations and through individual teachers pursuing higher degrees. In addition, students now graduate from preservice programs with a far greater knowledge and understanding of how children learn science and the theoretical bases of their own teaching than their experienced colleagues.

Nevertheless, preservice teacher education is in a tenuous position. While major government initiatives have impacted the status of preservice teacher education by relocation of programs to universities and a plethora of government inquiries and reports urging a greater attention to science education, the situation is constrained by shrinking educational funding. The reduction in grants may not stall research, but has the potential to limit and divert research from intensive field based studies to more economical action research studies involving undergraduate students.

The reduction in funding to universities will also increase the pressure on the quality of preservice teaching as previously suggested, both in terms of philosophical frameworks for pedagogy and availability of suitably qualified staff. These warning signs, evident in tight economic circumstances, have clear implications for science teacher preparation in other countries where economic rationalism is tending to subvert the needs of good teaching and research.

Another issue that remains problematic, without clear guidelines or intentions being signaled by key stakeholders, includes the impact of the competency movement on science teacher preparation. A competency framework has been developed for accrediting Australian teachers, but implementation is not high on the political agenda and the extent to which it will influence preservice education remains to be seen. Teacher education is an easy target, highly politicized and subject to constant review; consequently the development of science teacher education has become reactive to political decision making. The trend in Australia has been for governments to rely on reviews and reports that purport to reveal the current weaknesses with respect to education. The interpretation and implementation of these reports by governments and institutions are often made to suit their own purposes, thus leading to knee jerk reactions and inconsistencies in the development of programs.

A further unsettling feature of the political context for elementary science teacher preparation is a newly commissioned review of the tertiary education sector, reported to the Commonwealth government in March, 1998. Preliminary indications are that there will be major ramifications for funding and operational aspects of universities.

Despite all of the reviews, research, and curriculum redevelopment based on contemporary learning theory, there is little evidence to suggest that preservice science teacher education in Australia achieves much long term success in building graduates' self-confidence to teach science. While short term success has been demonstrated for some courses, there appear to be lasting effects in only a limited number of beginning teachers. We do not think this situation is unique to Australia, and suggest that the research agenda internationally could focus more on the social and environmental contexts in which elementary teachers try to teach science.

A positive and continuing feature of ongoing developments in preservice elementary science teacher education has been the role of ASERA. The work of this association of science education researchers may be a useful model for other countries attempting to cope with rapid change. As a collegial and supportive association that has mentored new researchers and provided a central forum for exchange of ideas over a number of years, it has been the only consistent voice for science teacher education. Participation in ASERA conferences has been a key form of professional development for elementary science teacher educators in providing a means to report research and engage in dialogue, consequently being the greatest agent for real positive change in elementary science teacher curriculums in preservice programs (Appleton & Symington, 1996). However, unlike professional teacher organizations, it does not have a permanent secretariat or organization that can lobby government or relevant industry groups. Its audience has therefore been limited to science education researchers who tend to be university-based. Other organizations such as ASTEC have been advocates, but their role as arms of government are limited, and subject to political restructuring of administration. The only other major player in supporting science education is the AAS, but its lack of a funding base limits its influence.

In conclusion, teaching as a profession is not highly valued in the Australian psyche. Cycles of teacher shortage and oversupply influence planning in teacher education and increase competitiveness between institutions. Issues of quality of students and attractiveness of a career in teaching are perennial concerns. Related to this, and especially problematic for science education, is the persistent occurrence of poor attitudes toward science and lack of personal commitment to teaching science found in undergraduate students in preservice elementary programs, and in the teaching force. These issues are not unique to Australia, but reinforce the need to highlight the important role of preservice elementary science teacher educators as advocates of science. They also raise the question of why we teach science to young children. While economic and technological purposes dominate the reasons given by governments, issues concerning personal understanding and the development of a scientific way of seeing the world are legitimate and may ultimately be more successful in fostering an understanding of science among young children.

Over the last thirty years, Australian preservice education has moved from essentially 1-year teacher training to a 4 or 5-year qualification. Many countries in the world are grappling with similar prospects. Amalgamation of teacher education institutions and upgrading of qualifications of staff is an international issue. Perhaps some of the experiences described here which Australia has sustained can provide insights to preservice elementary science teacher educators and administrators in these nations.

REFERENCES

Appleton, K. (1977). Is there a fairy godmother in the house? *Australian Science Teachers Journal, 23* (3), 37-42.

Appleton, K. (1983). Beginning student teachers opinions about teaching primary science. *Research in Science Education, 13*, 111-119.
Appleton, K. (1984). Student teachers opinions (A follow-up). *Research in Science Education, 14*, 157-166.
Appleton, K. (1991). Mature-age students - How are they different? *Research in Science Education, 21*, 1-9.
Appleton, K. (1993). Using theory to guide practice: Teaching science from a constructivist perspective. *School Science and Mathematics, 5*, 269-274.
Appleton, K. (1995). Student teachers confidence to teach science: Is more science knowledge necessary to improve self-confidence? *International Journal of Science Education, 17*, 357-369.
Appleton, K., & Kindt, I. (1997, July). *Beginning primary teachers and their teaching of science*. Paper presented at the annual conference of the Australasian Science Education Research Association, Adelaide, Australia.
Appleton, K., & Symington, D. (1996). Changes in primary science over the last decade: Implications for the research community. *Research in Science Education, 26*, 299-316.
Australian Academy of Science [AAS]. (1994). *Primary investigations*. Canberra, Australia: Author.
Bennett, N. (1984). Primary science in Australia and South Australia in 1995. In D. Symington (Ed.), *Primary science: The next ten years* (supplementary pp. 1-9). Malvern, Victoria, Australia: Victoria College.
Biddulph, F., & Osborne, R. (1984). *Making sense of our world: An interactive teaching approach*. Hamilton, New Zealand: SERU, University of Waikato.
Carmichael, L. (Chair). (1992). *The Australian vocational certificate training system*. Employment Skills Formation Council, Canberra, Australia: Australian Government Publishing Service.
Curriculum Corporation. (1994a). *A statement on science for Australian schools*. Carlton, Victoria, Australia: Author.
Curriculum Corporation. (1994b). *Science - A curriculum profile for Australian schools*. Carlton, Victoria, Australia: Author.
Dawkins, J.S. (1987). *Higher education: A policy discussion paper*. Canberra, Australia: Australian Government Publishing Service.
Department of Employment, Education and Training [DEET]. (1989). *Discipline review of teacher education in mathematics and science*. Canberra, Australia: Australian Government Publishing Service.
Dooley, J., & Lucas, K. (1981). Attitudes of student primary teachers towards science and science teaching. *Australian Science Teachers Journal, 27* (1), 77-80.
Farnsworth, I., & Jeans, B. (1994). Determinants of the competence and confidence of teacher education students studying primary science education. *Research in Science Education, 24*, 368-369.
Fensham, P.J. (1994). Progression in school science curriculum: A rational prospect or a chimera? *Research in Science Education, 24*, 76-82.
Finn, T. (Chair). (1991). *Young peoples participation in post compulsory education and training*. Report of the Australian Education Council Review Committee. Canberra, Australia: Australian Government Publishing Service.
Garnett, P., & Tobin, K. (1984). Reasoning patterns of preservice elementary and middle school science teachers. *Science Education, 68*, 621-631.

Ginns, I.S., & Foster, W.J. (1983). Preservice elementary teacher attitudes to science and science teaching. *Science Education, 67*, 277-282.

Ginns, I., & Watters, J. (1997, July). *Theory into practice for primary science education: An analysis of the stories of four beginning teachers*. Paper presented at the annual conference of the Australasian Science Education Research Association, Adelaide, Australia.

Ginns, I.S., Watters, J.J., Tulip, D.F., & Lucas, K.B. (1995). Changes in preservice elementary teachers sense of efficacy in teaching science. *School Science and Mathematics, 95*, 394-400.

Goodrum, D. (Ed.) (1993). *Science in the early years of schooling: An Australian perspective.* Key Centre Monograph No. 6. Perth, Australia: Key Centre for School Science and Mathematics, Curtin University of Technology.

Goodrum, D., Cousins, J., & Kinnear, A. (1992). The reluctant primary school teacher. *Research in Science Education, 22*, 163-169.

Hand, B., & Peterson, R. (1995). The development, trial and evaluation of a constructivist teaching and learning approach in a preservice science teacher education program. *Research in Science Education, 25*, 75-88.

Hardy, T., Bearlin, M., & Kirkwood, V. (1990). Outcomes of the Primary and Early Childhood Science and Technology Education Project at the University of Canberra. *Research in Science Education, 20*, 142-151.

Henry, J.A. (1977). Victorian primary science: Five years on. *Australian Science Teachers Journal, 23* (1), 49-55.

Hope, J., & Townsend, M. (1983). Student teachers understanding of science concepts. *Research in Science Education, 13*, 177-184.

Kirkwood, V., Bearlin, M., & Hardy, T. (1989). New approaches to the inservice education in science and technology of primary and early childhood teachers (Or Mum is not dumb after all). *Research in Science Education, 19*, 174-186.

Martin, L.H. (Chair). (1964). *Tertiary education in Australia.* Report of the Committee on the Future of Tertiary Education in Australia to the Australian Universities Commission, Vol.1. Canberra, Australia: Commonwealth Government Printer.

Mayer, E. (Chair). (1992). *Putting general education to work: The key competencies report.* Committee to Advise the AEC and MOVEET. Canberra, Australia: Sands McDougall.

Napper, I., & Crawford, G. (1990). Focus folklore: Reflections of focus teachers on the Sci-Tech In-service Project. *Research in Science Education, 20*, 230-239.

National Working Party on Teacher Competency Standards. (1996). *The national competency framework for beginning teaching.* Leichhardt, NSW, Australia: Australian Teaching Council.

Nikkerud, G., & D'Unienville, M. (1995, January). *Interactive science and the national statement and profile*. Paper presented at the national Conference for Primary Teachers and Educators, Canberra, ACT, Australia.

Osborne, R., & Freyberg, P. (1985). *Learning in science. The implications of children's science*. Auckland, New Zealand: Heinemann.

Owen, J.M., Johnson, N.J., & Welsh, R.J. (1985). *Primary concerns: A project on mathematics and science in primary teacher education*. Canberra, Australia: Commonwealth Tertiary Education Commission.

Skamp, K. (1987). Pre-service teachers: Process skill entry behaviour and opinions about teaching primary science. *Research in Science Education, 17*, 76-86.

Skamp, K. (1991). Primary science and technology: How confident are teachers? *Research in Science Education, 21*, 290-299.

Skamp, K., & Power, C. (1981). Primary science, inquiry and classroom demands: Pre-service teachers responses and perceptions. *Research in Science Education, 11*, 34-43.
Smith, R.G. (1997). Before teaching this I'd do a lot of reading. Preparing primary student teachers to teach science. *Research in Science Education, 27*, 141-154.
Symington, D. (1974). Why so little primary science? *Australian Science Teachers Journal, 20* (1), 57-62.
Symington, D. (1980). Primary school teachers knowledge of science and its effect on choice between alternative verbal behaviours. *Research in Science Education, 10*, 69-76.
Symington, D. (1982). Lack of background in science: Is it likely to always adversely affect the classroom performance of primary teachers in science lessons? *Research in Science Education, 12*, 64-70.
Symington, D., & Mackay, L. (1993). Response to the discipline review of teacher education in mathematics and science. *Research in Science Education, 23,* 286-292.
Varley, P.J. (1975). *Science in the primary school.* Brisbane, Australia: Research Branch, Department of Education, Queensland.
Watters, J.J., & Ginns, I.S. (1994). Self-efficacy and science anxiety among preservice primary teachers: Origins and remedies. *Research in Science Education, 24*, 348-357.
Whitehead, G., Symington, D., Mackay, L., & Vincent, A. (1993). *The influence of discipline reviews on higher education: Review of teacher education in mathematics & science.* Canberra, Australia: Australian Government Publishing Service.

Chapter 3

Reform in Science Teacher Education in Italy

The Case of Physics

Lidia Borghi, Anna De Ambrosis, and Paolo Mascheretti
University of Pavia, Italy

Abstract: In this chapter, we present the problem of the initial preparation of science teachers in Italy. We describe prospective changes both for the preparation of primary school and secondary school teachers and compare with the present situation. We draw attention, in particular, to the preparation of physics teachers as an example of science teacher preparation. Here we report the complete path of their studies (undergraduate and graduate) and discuss the main features that graduate courses should have to prepare effectively preservice teachers for their future work.

At the present time, education in science is accepted as crucial for the cultural development of individuals. The improvement of science teacher preparation is a common aim in different countries. Consistent efforts in this direction are making progress in Italy: changes in teacher preparation have been established by law and a reform of the whole school system is now under discussion. This chapter aims at presenting these changes.

In order to provide a frame, a brief presentation of the Italian school system is initially given. Next the current teacher education system is described. Third we present innovations in teacher preparation for primary and secondary school, focusing on the main features of the new undergraduate and graduate courses respectively for primary school teachers and high school teachers. The case of physics is considered in detail, by drawing attention to the fundamental criteria on which the organization of courses and methodological choices are grounded. We also show how considerations about physics teacher preparation can be extended to other scientific disciplines.

S.K. Abell (ed.), Science Teacher Education, 31–43.

THE PRESENT ITALIAN SCHOOL SYSTEM

Children enter compulsory school at six years (generally after three years of preschool), and complete it after five years of primary school and three years of junior secondary school. (A law that deeply modifies the organization of our school system, establishing compulsory schooling from 5 to 15 years of age, is now under discussion and it is expected to be approved in the near future).

Compulsory schooling (see Table 1) is the same for all children, including those with mental or physical disabilities for whom specialized support is provided.

Table 1. The Present Italian School System

School	Duration (Years)	Starting Age (Years)
Preschool	3	3
Primary school*	5	6
Junior secondary school*	3	11
High school	5	14
University	4 (5-6)	19

* indicates compulsory schooling

According to the national syllabus introduced in 1985, science should be taught from the very beginning of primary school without a separation of scientific disciplines. Presently teaching activity is shared by three teachers, one of whom has the responsibility for science and mathematics education. In junior secondary school (which is compulsory and lasts three years), the number of teachers increases to eight, but mathematics and science are still taught by the same teacher.

High school (non-compulsory) lasts five years and consists of different sections: Liceo Scientifico (oriented to science and mathematics), Liceo Classico (oriented to the humanities), Liceo Artistico (oriented to the fine arts), and a number of Technical Institutes (Commercial, Industrial, Agricultural, Construction, etc.). As in compulsory school, the syllabus is established at the national level. Scientific disciplines are taught by specialist teachers. In particular, mathematics and physics can be taught by the same teacher; biology, chemistry and health science are taught by another teacher. At the end of high school, students have to pass an examination designed by a national committee for each section of high school. After this examination, students may start state university without passing an entrance examination. (For a few disciplines the enrollment is limited and a selection of students is made by means of tests. The total number of Italian state universities is 64; the number of private universities is quite low). The preparation of students who enter the university is generally based more on humanities than on scientific subjects; a high frequency of drop-out is common, especially for undergraduate students in scientific disciplines.

PREPARATION OF TEACHERS

In order to show how deeply the preparation of teachers will change in the near future, the present teacher education system is briefly illustrated. Since the preparation of teachers strongly depends on the school teaching level, the case of primary school teachers is described separately from that of secondary school teachers.

The Present Situation

Primary School Teachers

According to a long tradition, primary school teachers are prepared in a special section of high school (Istituto Magistrale). They do not earn a university degree and they usually receive a preparation oriented to methodology rather than to disciplines, especially in science (Borghi, De Ambrosis, and Massara, 1991). The seriousness of their limited education in science was emphasized when the new national curriculum for Italian primary school was introduced. It required that science education be an essential part of primary school from the very beginning and, in accordance with research findings, the science curriculum be experience-oriented (for example, see Bazzini et al., 1985; Bonera, Borghi, De Ambrosis, and Massara, 1983; Bonera, Castellani Bisi, Borghi, De Ambrosis, and Massara, 1981; Goldberg & Boulanger, 1981; Karplus & Their, 1970; McDermott, 1976, 1990a).

A consistent effort to enhance primary teachers' background in science and to equip them with the tools necessary to improve their skills in science teaching has been carried out by the Italian Ministry of Education by launching a national plan of professional development for inservice elementary school teachers lasting five years. About 300,000 primary school teachers were engaged in compulsory courses (aimed at preparing teachers to implement the new curriculum in the classroom) in the following areas: art, languages, history and geography, mathematics, music, and science. The implementation of the national plan of training allowed the research groups in science education in a few Italian universities to test the effectiveness of their research on teacher preparation in proposing new models of inservice education (Borghi, De Ambrosis, and Massara, 1993). In particular, our experience in a number of courses confirmed that, inside the field of science education for teachers, physics can play a fundamental role, because it allows a complete path from the experiential phase to the formal one. This is to say that teachers must have a good knowledge of fundamental concepts of physics, and have direct experience of what a physics experiment is. We considered it necessary to help primary school teachers acquire significant experience in working with equipment, in recognizing the essential variables in an experiment, in testing models, etc.

The work carried out with inservice teachers suggested guidelines for designing the new project for initial university preparation of primary school teachers which we describe later in this chapter.

Secondary School Teachers

Teachers of scientific disciplines are prepared at the university where they earn one of the traditional science degrees: Biology, Chemistry, Geology, Mathematics, or Physics. For most disciplines, teaching-oriented courses are not offered and, even when such courses are available (as is the case of physics in a limited number of universities) they are not compulsory. In order to be admitted to high school teaching in public schools, besides having a degree in the discipline, passing a national qualifying examination is required. Teaching positions are conferred through a national competitive examination.

The instruction in the subject matter provided by university courses is sound but specialized in each discipline. It fits with the separation of scientific areas in high school, but it is generally too narrow to cover the range of scientific disciplines in junior high school. In any case, teaching as an activity of mediation between discipline and knowledge building of students is usually disregarded. Practicing teachers usually try to fill this lack by attending professional development courses, often funded by the Ministry of Education and organized by different public institutions such as: their schools, teacher associations, Regional Institutes of Educational Research (IRRSAE) and the Ministry. These courses are not compulsory and do not have common features. Based on our experience, teachers appreciate courses grounded in laboratory activity that give them the opportunity to develop and test experiments to include in their teaching practice.

A national plan, National Plan for Computer Science, for inservice teacher education was organized, starting in 1985, with the aim of introducing the use of computers to improve learning in different disciplines, in particular in mathematics and physics. These courses have given inservice teachers the opportunity of participating in the debate on the use of new technologies in science teaching (Arons, 1984, 1990; Bacon, 1992; Hewson,

1985; Hicks & Laue, 1989; McDermott, 1990b; Schwartz, 1989; Taylor, 1987, 1988).

Universities also provide specialized courses on different subjects. The involvement of university researchers in science education in a number of courses allows teachers to gain awareness of the existing research on teaching and learning while providing researchers with new insights on teachers' needs and problems.

The Prospective Situation

A recent law assigns the universities a fundamental role in the education of teachers at every school level. It establishes major changes in initial teacher preparation both for primary and secondary school.

Primary School Teachers

According to the new law, starting from the academic year 1998-99, people who want to become primary-school or preschool teachers, must earn a university degree (4-year) by following a specific curriculum, including courses in content and professional education. The law establishes general criteria and guidelines for the preparation of teachers and gives universities the complete responsibility for organizing the courses. In particular, teachers are expected to attend courses for eight 6-month periods (2,000-2,400 hours total). The percentages of time devoted respectively to disciplinary, general education, laboratory, and teaching practice, must be at least 35%, 20%, 10%, 20%. Professional courses cover 5% and the remaining time can be devoted to elective courses. In the first two years, courses are common to preschool and primary preservice teachers.

As an example, in Table 2 we report the proposal of a committee of the Ministry of Education (MPI) and the Ministry of University and Scientific and Technological Research (MURST). The total number of semester courses to be attended in four years is 50, including eight courses for teaching practice and six elective courses.

As Table 2 shows, 17 courses are devoted to content matter in different areas and are accompanied by teaching practice. A preservice teacher is expected to delve into two of these areas, with not more than one selected from arts, music, and physical education.

A thesis is required and its discussion is part of the final examination. Courses and thesis work aim at developing competence both in subject matter and in cognitive and methodological aspects of teaching. They offer the possibility of building a bridge between psycho-pedagogist and subject matter competencies by studying teaching methods in the context in which they are to be implemented

Table 2. Courses for Primary School Preservice Teachers: A Proposed Set of Courses for Primary Preservice Teachers in Italy

Subject Areas	Number of Semester Long Courses	Teaching Practice (6-month units)
Italian language	3	2
Foreign language	2	2
Mathematics	2	2
Science	4	2
History, geography, and social studies	3	2
Arts	1	2
Music	1	2
Physical education	1	2
Sociology, anthropology, psychology, and pedagogy	13	
Philosophy	2	
Hygiene and sanitation	2	
Law and organization of school	2	

Secondary School Teachers

According to the new law, those who want to become secondary school teachers (either in junior high school or in high school) must attend graduate school, after receiving an undergraduate degree in a discipline. The 2-year graduate school is to be organized by disciplines: Natural Sciences, Physics-Mathematics-Computer Science, Humanities, Language and Literature, Foreign Languages, Economy and Law, Arts, Music and Media, Hygiene and Sanitation, Technology, Physical Education. For each section, courses cover a total of at least 1,000 hours, including teaching practice. The percentages of time devoted respectively to disciplinary knowledge, general education and professional, laboratory, apprenticeship, are at least 20%, 20%, 20%, 30%.

A committee of MPI and of MURST proposed an example of organization of the graduate school according to which courses (of 50 hours minimum) are divided as follows: six common courses in education, six courses in specific subjects, two courses for teaching practice, and electives. A thesis is part of the final examination.

Common courses deal with general issues and are attended by preservice teachers of different sections. Courses on specific subjects must include at least one devoted to the history of the discipline and three to a laboratory on the teaching of the discipline. Particular attention is given to teaching practice, which is organized by the university in collaboration with inservice teachers and local authorities. The graduate school complements the preparation obtained in the university studies. Its innovative character consists of the professional preparation of teachers (hitherto completely disregarded), which accompanies a sound disciplinary preparation.

In order to show an example of the complete curriculum for secondary teacher preparation, the case of physics is considered here in detail. The situation in the other sciences is quite similar.

THE CASE OF PHYSICS TEACHERS

As already mentioned, a student who wants to become physics teacher must obtain an undergraduate physics degree, attending university courses for at least four years. In the first three years, the courses are the same for all physics students. In the fourth year, the courses are differentiated by physics subfield; for example, at the University of Pavia there are six different sections: Nuclear and Subnuclear Physics, Physics of Matter, Astrophysics and Space Physics, Applied Physics, Theoretical Physics, Physics Education, and History of Physics.

The complete curriculum for the section "Physics Education and History of Physics" (Didattica e Storia della Fisica), which is for students who want to become secondary school teachers, is outlined in Table 3. (In the list we use English titles for courses, followed, in brackets, by the official Italian names.)

While in the first three years courses are compulsory and common to every section, in the fourth year students are allowed to choose the courses numbered 17, 18, and 19 from a broad list of specialized courses, some of them oriented to teaching (such as Foundations of Physics, New Technologies in Education, and Physics Teaching), and others to specific physics topics (for example Astronomy, Electromagnetic Waves, Optics, Nuclear Physics, Material Physics, Radioactivity, Relativity, and Solid State Physics). At the end, each student carries out a thesis work which is presented and discussed in the final examination. Knowledge of English is required; foreign language courses are usually taken within the first two years.

According to the new law, after graduating in physics, preservice teachers must attend graduate school to obtain the teaching license. The courses of the graduate school start in the academic year 1999-2000. Following the guidelines of the MPI and MURST, each university is planning the implementation of the graduate school. A significant example

Table 3. The Undergraduate Physics Teaching Degree

First year
1. Physics I (Fisica Generale)
2. Experimental Physics I (Esperimenti di Fisica I)
3. Calculus I (Analisi Matematica I)
4. Geometry (Geometria)
Second year
5. Physics II (Fisica Generale II)
6. Experimental Physics II (Esperimentazioni di Fisica II)
7. Calculus II (Analisi Matematica II)
8. Chemistry (Chimica)
9. Classical Mechanics (Meccanica Razionale con elementi di Meccanica Statistica)
Third year
10. Mathematical Methods of Physics (Metodi Matematici della Fisica)
11. Theoretical Physics (Istituzioni di Fisica Teorica)
12. Experimental Physics III (Esperimentazioni di Fisica III)
13. Structure of Matter (Struttura della Materia)
14. Nuclear and Sub nuclear Physics (Istituzioni di Fisica Nucleare e Subnucleare)
Fourth year
15. A course among:
Complements to Introductory Physics (Complementi di Fisica)
Advanced Physics (Fisica Superiore)
History of Physics (Storia della Fisica)
16. A course between:
Teaching Tools and Methods (Preparazione di Esperienze Didattiche)
Laboratory of instrumental Physics (Laboratoria di strumentazioni fisiche)
17. Annual course (selected from complete course offerings in Physics)
18. Semester long course (selected from complete course offerings in Physics)
19. Semester long course (selected from complete course offerings in Physics)

of such a program has been designed at the University of Bologna (Frabboni, Tomasini Grimellini, Manini, and Pellandra, 1994) (see Table 4).

According to the new law, after graduating in physics, preservice teachers must attend graduate school to obtain the teaching license. The courses of the graduate school start in the academic year 1999-2000. Following the guidelines of the MPI and MURST, each university is planning the implementation of the graduate school. A significant example of such a program has been designed at the University of Bologna (Frabboni, Tomasini Grimellini, Manini, and Pellandra, 1994) (see Table 4).

The physics teaching courses will be offered in university physics departments because of the fundamental choice made by Italian researchers

Table 4. Proposed Graduate Program for Physics Teaching, University of Bologna

Common courses:

- Pedagogy and History of Education Systems (Pedagogia e Storia dei sistemi formativi)
- Methodology and Education (Metodologia e Didattica Generale)
- Sociology of Education (Sociologia dell'Educazione)
- Psychology (Psicologia)
- Cultural Anthropology (Antropologia culturale)
- School Organization and Law (Normativa legislativa e Cultura organizzativa

Courses on specific subjects (Physics)

- Epistemology and History of Physics I (Epistemologia e Storia della Fisica I)
- Epistemology and History of Physics II (Epistemologia e Storia della Fisica II)
- Physics Teaching I (Didattica della Fisica I)
- Physics Teaching II (Didattica della Fisica II)
- Laboratory of Physics Teaching I (Laboratorio di Didattica della Fisica I)
- Laboratory of Physics Teaching II (Laboratorio di Didattica della Fisica II)

Courses for teaching practice

- 2 semester long courses of teaching practice in schools

Subsidiary courses

- Preservice teachers freely choose two courses among the ones offered in the School.

in physics education to carry out their research inside physics departments (the same is true for the other disciplines). As a consequence, cooperation of different departments is required for success of the graduate program. In particular, a strong collaboration between pedagogy and physics experts must be kept in each university.

In our department at the University of Pavia, we are planning to offer the courses listed in Table 4 and additional courses on new technologies in education. In order to enhance student choice, a modular structure will be adopted. The main features of the courses are briefly described below.

The subject matter courses will take into account the need for physics to be taught in such a way that high school students perceive the cultural value of physics and acquire the capacity to frame and define new problems, instead of learning to solve only problems already defined and categorized.

Courses in Epistemology and History of Physics are meant to provide preservice teachers with a knowledge of the subject wide enough to allow them to incorporate the history of physics in their teaching. Research has shown how history of physics can improve student understanding of physics topics (Bevilacqua, Bonera, Borghi, De Ambrosis, and Massara, 1990; Matthews, 1992). However, teachers generally are not aware of this connection and they are not provided with examples of historical-based treatment of physics topics (Osborne, 1997).

Courses and laboratories in physics teaching will address the problems of knowledge transmission in teaching physics. The constructive character of learning and its consequences on knowledge acquisition should be evidenced (for example, Laws, 1997; Novodvorsky, 1997; Redish, 1994). In this frame, future teachers will play a role of mediator and guide in the classroom. In particular, they should acquire competence in understanding and discussing ideas and conceptions of students, in helping them to connect these with the concepts accepted by the physics community, in learning how to use their mathematical and analytical skills for a formal description of phenomena, and in synthesizing the knowledge acquired to gain both security and confidence in science and technology. Having in mind both the cultural and the professional needs of physics teachers, courses will be different from the traditional university courses, to avoid future teachers receiving knowledge rather than being helped to generate it.

In these courses preservice teachers must have the opportunity to carry out lab activities and autonomous investigations, so that they personally experience the kind of teaching they should promote in school. In summary, future teachers should be provided with opportunities to:

- Know the results of research in physics education.
- Be personally involved in experimental activity.
- Become familiar with lab equipment and teaching tools available in the schools, in particular become proficient with personal computers.
- Use new technologies which in the near future will be available in the schools (a significant example is MBL-Microcomputer-Based Laboratory-tools [Redish, Saul, and Steinberg, 1997; Thornton & Sokoloff, 1990]).
- Learn how to explore the ideas of their students and design instructional strategies to help students overcome difficulties.
- Analyze and discuss teaching strategies and new approaches.
- Appreciate working in groups and reflecting on their action in classrooms.
- Frame, in an historic dimension, the elaboration of physical theories and, in significant case studies, analyze the debate which accompanied their development.
- Learn how to provide alternative ways of thinking about a concept and pauses for interpretation and reflection.
- Design instructional strategies that help pupils experiment in different areas the passage from their everyday experience to representations of formal physics.

- Learn to integrate into their teaching activity different tools and approaches, discussion of everyday experience, use of simple experiments with easily available material and more complex equipment, together with multimedia tools (Borghi, De Ambrosis, Invernizzi, and Mascheretti, 1996).

Effective teaching, in fact, depends on the teacher's mastery of a wide range of strategies which maximize motivation and learning of pupils (Osborne, 1996). Research on teaching and learning in science must sustain the implementation of the graduate school in the same way as it showed the need for innovation in science teacher preparation.

CONCLUSION

Despite the high interest in science teacher education reform in Italy and its innovative spirit, a variety of obstacles prevent its implementation. University teachers are usually prepared to give traditional courses in different disciplines, and these courses do not provide the preparation that teachers need. Only in a limited number of universities is it possible to find researchers in disciplinary education who can support the learning of pedagogy together with content.

In order to improve the present situation, it seems essential to enhance in each university the interest in educational research so that sections of disciplinary curricula, such as science, be devoted to teacher preparation. Financial support for research in education is limited; at present, in Italy only 0.1% of resources for research are devoted to research in education, whilst in other European countries the amount is somewhat higher (for instance, in Germany the percentage is 1%).

At present, the number of disciplinary researchers working in science departments whose main research interest is science education, and of researchers in pedagogy and psychology, is lower than required for an effective implementation of the reform. Only 17 Italian university research groups in physics education exist (and physics is a positive exception), with a low number of researchers who, despite career difficulties and scarce financial support, carry out research in a fruitful collaboration, with good coordination at the national level and satisfactory results (Commissione Didattica del CNR 1991, 1992; National Project for Physics Education in High School, 1997).

A serious impulse to accelerate these changes in teacher education may come from the reform of the Italian school system. These reform measures are expected to be approved soon and, quite naturally, will draw attention to concomitant reforms in teacher preparation.

REFERENCES

Arons A.B. (1984). Computer-based instructional dialogs in science courses. *Science, 224*, 1051-1056.

Arons A.B. (1990). *A guide to introductory physics teaching*. New York, NY: John Wiley & Sons, Inc.

Bacon, R. A. (1992). The use of computers in the teaching of physics. *Computers and Education, 19*, 57-64.

Bazzini, L., Borghi, L., De Ambrosis, A., Ferrari, M., Massara, I., Mosconi Bernardini, P., Trivelli Ricci, P., & Vittadini Zorzoli, M. (1985). A project about the teaching of mathematics and science in the first two grades of primary school. *European Journal of Science Education, 7*, 29-36.

Bevilacqua, F., Bonera, G., Borghi, L., De Ambrosis, A., & Massara, C.I. (1990). Computer simulation and historical experiments. *European Journal of Physics, 11*, 15-24.

Bonera, G., Castellani Bisi, C., Borghi, L., De Ambrosis, A., & Massara I. (1981). Teaching science in elementary school: A research programme by the University of Pavia-Italy. *European Journal of Science Education, 4*, 479-480.

Bonera, G., Borghi, L., De Ambrosis, A., & Massara, I. (1983). Science education in the primary school: The problem of the initial training of teachers in Italy. *European Journal of Science Education, 2*, 141-145.

Borghi, L., De Ambrosis, A., & Massara, C.I. (1991). Physics education in science training of primary school teachers. *European Journal of Teacher Education, 14*, 57-63.

Borghi, L., De Ambrosis, A., & Massara, C.I. (1993). Inservice training of primary school teachers: An example with the use of computers. *European Journal of Teacher Education, 16*, 215-223.

Borghi, L., De Ambrosis, A., Invernizzi, C., & Mascheretti, P. (1996). Un modèle por la compréhension des propriétés des liquides. *Didaskalia, 8*, 139-153.

Commissione Didattica del CNR. (1991). *Per una educazione scientifica di base*. Pavia, Italy: La Goliardica Pavese.

Commissione Didattica del CNR. (1992). *Applicazioni dell'elaboratore nella didattica della fisica*. Napoli, Italy: CUEN.

Frabboni, F., Tomasini Grimellini, N., Manini, M. & Pellandra, C. (Eds.) (1994). *Scuola di specializzazione all'insegnamento secondario*, Bologna, Italy: CLUEB.

Goldberg, H. S., & Boulanger, F.D, (1981). Science for elementary school teachers: A quantitative approach. *American Journal of Physics, 49*, 120-124.

Hewson, P. H. (1985). Diagnosis and remediation of an alternative conception of velocity [illegible]

Hicks, R. B., & Laue, H. (1989). A computer assisted approach to learning physics concepts. *American Journal of Physics, 57*, 807-811.

Karplus, R., & Thier, H.D. (1970). *A new look at elementary school science*. New York: Rand McNally and Company.

Laws, P.W. (1997). Millikan lecture: Promoting active learning based on physics education research in introductory physics courses. *American Journal of Physics, 65*, 14-21.

Matthews M. (1992). Teaching about air pressure: A role for history and philosophy in science teaching. In S. Hills (Ed.), *The history and philosophy of science in science education* (Vol. II). (pp.121- 133). Kingston Ontario: Queen's University.

McDermott, L. (1976). Teacher education and the implementation of elementary science curricula. *American Journal of Physics, 44*, 434-441.

McDermott, L.C. (1990a) A perspective on teacher preparation in physics and other sciences: The need for special science courses for teachers. *American Journal of Physics, 58*, 734-742.

McDermott, L. (1990b). Research and computer-based instruction: Opportunity for interaction. *American Journal of Physics, 58*, 452-462.

National Project for Physics Education in High School. (1997) *Oscillatori e Oscillazioni*, Bologna, Italy: Italian Physics Society.

Novodvorsky, I. (1997). Constructing a deeper understanding. *The Physics Teacher, 35*, 242-245.

Osborne, J. (1996). Untying the Gordian knot: Diminishing the role of practical work. *Physics Education, 31*, 271-278.

Osborne, J. (1997). Placing the history and philosophy of science on the curriculum: A model for the development of pedagogy. *Science Education, 81*, 405-423.

Redish, E.F. (1994). Implications of cognitive studies for teaching physics. *American Journal of Physics, 62*, 796-803.

Redish, E.F., Saul, J. M., & Steinberg, R.N. (1997). On the effectiveness of active-engagement microcomputer-based laboratories. *American Journal of Physics, 65*, 45-54.

Schwartz, J. L. (1989). Symposium: Visions for the use of computers in classroom instruction. *Harvard Educational Review, 59*, 50-61.

Taylor, E. F. (1987). Comparison of different uses of computers in Teaching Physics. *Physics Education, 22*, 202-211.

Taylor, E. F. (1988). Learning from computers about physics teaching. *American Journal of Physics, 56*, 975-980.

Thornton, R K., & Sokoloff D. (1990). Learning motion concepts using real-time micro-computer based laboratory tools. *American Journal of Physics, 58* , 858-867.

Chapter 4

Science Teacher Preparation in Lebanon

Reality and Future Directions

Saouma BouJaoude
American University of Beirut, Lebanon

Abstract: The purpose of this study was to answer the following questions: (a) What are the theoretical perspectives driving science teacher preparation programs in Lebanon? (b) What are the requirements of Lebanese science teacher preparation programs? and (c) What are the similarities and differences among the variety of science teacher preparation programs offered in Lebanon? Data sources for this study included: official governmental documents and mandates related to teacher preparation; institutional catalogues and syllabi of courses; and interviews with science education professors. Results of the study showed that teacher preparation programs in Lebanon are characterized by: (a) post-graduate programs that prepare secondary teachers with significant amount of science background; (b) three and 4-year programs that prepare elementary classroom teachers or science/mathematics teachers; (c) the absence of university level programs for the preparation of intermediate school science teachers; (d) the requirement of a thesis in many of the programs; (e) a lack of emphasis on field work; and (f) the adoption of an orientation that has some characteristics of the academic and technological orientations to teacher preparation.

Education in Lebanon has a special flavor. The freedom of education guaranteed by the Lebanese constitution has allowed private schools, universities, and colleges to flourish. These institutions are affiliated with international and national religious, independent non-profit, and independent for-profit organizations. Presently, at the tertiary level, there are 19 private universities or colleges, nine of which offer science teacher education programs. Also, several 2-year private colleges are involved in science teacher preparation. Likewise, the Lebanese government is involved in science teacher preparation through the Lebanese University, the Center for Educational Research and Development (CERD), and the Ministry of Technical and Vocational Education.

S.K. Abell (ed.), Science Teacher Education, 45–74.

Colleges and universities in Lebanon can be classified into four categories based on the higher education model they follow: American, French, Arab, or Lebanese. The only university following an Arab model of higher education (The Arab University) does not offer science teacher education programs as of yet. The universities that follow an American model or a French model have programs patterned after similar ones in universities in the US or in France (Freiha, 1997).

The French model is different from the American model in that it is organized by years rather than by courses. In addition, programs in institutions that follow the French model are characterized by early specialization, absence of a liberal arts core, and lack of electives. The Lebanese model, however, has its own distinct character with ideas derived from more than one of the other models.

The preparation of science teachers at the elementary, middle school, and high school levels is integral to the missions of private as well as public Lebanese institutions of higher education. Competition among the institutions has created a wide variety of science teacher preparation programs, each with its own theoretical perspective, set of requirements, and characteristics, but each preparing science teachers for Lebanese schools and in some cases for schools in the Arab region. These institutions prepare science teachers in a variety of programs and institutional structural units and offer different types of degrees. Table 1 presents the names of Lebanese colleges and universities that offer education programs along with the degrees they offer.

To understand the variety of teacher preparation programs offered by universities in Lebanon requires an understanding of the Lebanese pre-college educational ladder. In 1967, Lebanese Law Number 9099 instituted four stages in the pre-college educational system: preschool, elementary, intermediate, and secondary. Law Number 10227 (1997) maintained these four stages but refined the number of years required at each. The preschool stage consists of two years, the elementary stage consists of six years divided into two [illegible] cycles, and the intermediate and secondary levels consist of three years each, for a total of 14 years.

Lebanese students are required to follow the Lebanese curriculum. As Figure 1 demonstrates, this curriculum is common for all students until Grade 10. In Grade 11 students may choose to follow the humanities stream or the science stream. Those who choose the humanities stream may choose to continue with the humanities and literature stream or follow the social sciences and economics stream in Grade 12. The students who choose the science stream in Grade 11 level may choose the general sciences stream or the life sciences stream in Grade 12. Each stream consists of a fixed number of subjects that all students who choose the stream are required to follow.

Table 1. Lebanese Colleges and Universities that Prepare Science Teachers, Degrees They Offer, and Number of Years of Study

Institution	Degree*	Duration
American University of Beirut	BA	3 years
	Teaching Diploma	1 year
Haigazian University	BA	3 years
	Normal Diploma	1 year
Higher College for Teacher Preparation	Education License	4 years
Lebanese University	Education License	4 years
	Diploma of Higher Studies	2 years
	Certificate of Qualification	2 years
Lebanese American University	BA	3 years
	Teaching Diploma	1 year
Middle East College	BA	3 years
Notre Dame University	Teaching Diploma	1 year
University of Saint Joseph	Education License	4 years
University of Balamand	Education License	1 year
	Teaching Diploma	4 years
University of the Holy Spirit in Kaslik	Education License	4 years
	University License	3 years

*The Teaching Diploma, Normal Diploma, Diploma in Higher Studies, and Certificate of Qualification require an undergraduate science degree. Admission to universities requires the Lebanese Baccalaureate. Lebanese students holding the Baccalaureate are admitted as sophomores in universities that follow an American model.

There is no possibility of elective courses within the stream. All students take science at all levels. However, the number of periods per week varies with the level and stream the student selects. Table 2 presents the number of periods of science at each grade level and in each stream. The "General Science" designation refers to courses that include life, physical, and Earth science. The "Science" designation at the Grade 11 level refers to the name

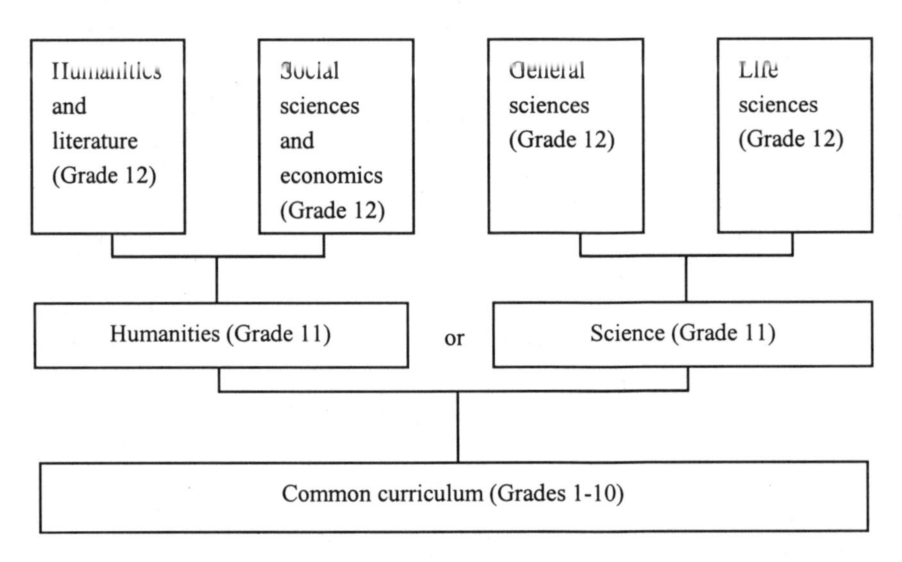

Figure 1. Structure of the Lebanese Educational Ladder

of the stream, as described above. Note that the 1.5 periods of chemistry and physics at Grade 7 represents a bureaucratic compromise, typically solved by having the same teacher for both subjects; the teacher then devotes the required time to each course.

Table 2. Number of Periods per Week of General Science, Biology, Chemistry, and Biology Taught at Each Grade Level of the Lebanese Educational System

Grade	1	2	3	4	5	6	7	8	9	10	11		12			
											S	H	GS	L	SS	H
General Science	2	2	3	4	4	5										
Biology							3	2	2	2	2			6		
Chemistry							1.5	2	2	2	3		4	5		
Physics							1.5	2	2	3	5		7	5		
Science Literacy												3			4	3
Total	2	2	3	4	4	5	6	6	6	7	10	3	11	16	4	3

S = Science, H = Humanities, GS = General Sciences, L = Life Sciences,
SS = Social Sciences and Economics

Lebanese students sit for official national examinations at the end of the intermediate and the secondary stages. The official examination taken at the end of the intermediate stage is common to all students in the general education[1] system. However, the examination taken at the end of the secondary education stage, called the Lebanese Baccalaureate and required for admission to universities, is divided into four different sections: humanities and literature, social sciences and economics, general sciences, life sciences.

Lebanon is presently in the midst of educational reform. The Lebanese Government has enacted a new educational ladder and CERD is preoccupied with developing new curricula for all subject areas at all pre-college levels (CERD, 1995; Public Law Number 10227, 1997). As a result, it is essential to understand the current status of science teacher preparation. Additionally, sharing ideas about the status and future directions of science teacher preparation in Lebanon with an international audience may help create a dialogue about teacher preparation worldwide. Consequently, the purpose of this study was to answer the following questions:

1. What are the theoretical perspectives driving science teacher preparation in Lebanon?
2. What are the requirements of Lebanese science teacher preparation programs?
3. What are the similarities and differences among the Lebanese science teacher preparation programs?
4. What policies drive Lebanese science teacher preparation programs?
5. What are the future directions of science teacher preparation in Lebanon?

LITERATURE REVIEW

Research studies have focused on the structural components of Lebanese teacher education programs without neglecting conceptual components. Most of these studies investigated teacher education in general with only a few focusing on science teacher education. The most recent studies by Farah-Sarkis (1997) and Freiha (1997) found that most programs emphasize theoretical rather than practical issues. Moreover, Freiha found that the nature of these programs was influenced by the model of higher education espoused at the institution in which they were offered. Consequently, programs at American style institutions were similar to programs offered at American universities while those at French style institutions were patterned after similar ones in French universities. Farah-Sarkis, on the other hand, found that there was no balance between theoretical and practical components of the Lebanese teacher education programs, with the percentage of time dedicated to practical work ranging from 5.2% to 25%.

[1] There are two parallel systems of education in Lebanon: General and Technical.

Wehbe (1984) found that teacher preparation at Lebanese colleges, universities, and specialized institutes emphasizes technical and theoretical issues and neglects moral and ethical components of the teaching-learning process. Murr (1983), on the other hand, found that the major difference between the elementary and intermediate teacher education programs implemented by CERD is that elementary teachers are prepared as classroom teachers while those at the intermediate level are prepared as specialized subject matter teachers (secondary programs were not examined).

Eido (1983) described the goals and structures of the teacher preparation programs offered at the Higher Institute for Teacher Preparation, a private institution owned by Almakassed Benevolent Islamic Association. According to him, the Institute offered two types of programs, one for elementary level classroom teachers, and another for intermediate level science and mathematics teachers. One distinguishing feature of the preparation of intermediate level teachers was that some of them were prepared to teach both mathematics and science in Grades 6 and 7 while others were prepared to teach either science or mathematics in Grades 8 and 9. Haddad (1983) described the goals and structures of the teaching diplomas in elementary and secondary (including intermediate) education offered at the American University of Beirut. According to Haddad, teachers working toward a Teaching Diploma in elementary or secondary education followed a program consisting of four components: pre-requisite subject matter courses, general pedagogy courses, methods courses, and field work. The difference between elementary and intermediate teachers, though, was in the number of subject matter courses taken and in the areas taught; while elementary teachers were prepared to teach both science and mathematics, intermediate school teachers were prepared to teach either science or mathematics.

THEORETICAL FRAMEWORK

Without agreement on the ideal theoretical perspective from which to study teacher preparation (Anderson & Mitchener, 1994), two convincing frameworks that can be employed in this endeavor include those of Feiman-Nemser (1990) and Kennedy (1990). According to Feiman-Nemser, researchers discussing teacher preparation have focused on either structural or conceptual issues. Structural issues include the general organization of programs such as the number of years to complete a program, the number of required credit hours of education and content, the duration of field-based experience, and alternative certification methods. Conceptual issues, on the other hand, include the different views of teaching and theories of learning to teach that drive programs. Conceptual orientations can be academic, practical, technological, personal, and critical/social, with reflective teaching being a professional stance that can be emphasized in any of these orientations (Feiman-Nemser, 1990). According to Feiman-Nemser, the academic orientation is "primarily concerned with the transmission of knowledge and the development of understanding" (p. 221). The practical orientation "focuses on the elements of craft, technique and artistry that skillful practitioners reveal in their work" and on "the primacy of experience as a source of knowledge about teaching" (p. 222). The technological orientation:

> Focuses attention on the knowledge and skills of teaching. The primary goal is to prepare teachers who can carry out the task of teaching with proficiency. Learning to teach involves the acquisition of principles derived from the scientific study of teaching. (p. 223)

The personal orientation "places the teacher-learner at the center of the educational process . . . The teacher's own personal development is a central part of teaching" (p. 225). Finally, the critical/social orientation portrays an optimistic faith in schools as agents of social change and preservers of social inequalities.

Kennedy (1990) suggested that teacher preparation programs can be classified as either emphasizing teaching students a large body of knowledge or preparing them to think and use problem solving strategies to analyze and learn from new situations. These two perspectives, according to Kennedy, are useful frameworks that can be used to study teacher preparation programs.

METHOD

Data sources for the study included: (a) official governmental documents, public laws, decrees, and mandates related to teacher preparation; (b) institutional catalogues or brochures describing science teacher preparation programs in all Lebanese institutions; (c) syllabi of courses offered in each of the science teacher preparation programs; and (d) interviews with university and program administrators and science education professors. Interviews were conducted with seven chairpersons, twelve professors, two registrars, one director, and two academic deans. In addition, interviews were conducted with the administrators of three 2-year institutions and the director of in-service and preservice teacher education at CERD. Data from each of the institutions was analyzed to identify the structural components as well as the conceptual orientations of science teacher preparation in each institution using the framework provided by Feiman-Nemser (1990). The general organization of the programs such as the number of years to complete a program, the number of required credit hours of education and content, and the duration of field-based experience were compared. Likewise, the programs were analyzed in terms of their conceptual orientations that can be academic, practical, technological, personal, or critical/social (Feiman-Nemser, 1990). Decisions regarding program orientations were based on an analysis of the characteristics of each of the programs, as revealed in the documents and the interviews, and a comparison of the characteristics to those identified by Feiman-Nemser. The results of the analysis of each institution were compared to identify patterns across institutions.

RESULTS

The results are presented in three sections. The first section presents the theoretical perspectives, requirements, and policies driving the education of public school science teachers. The second section presents the same information about the education of private school science teachers. Finally, the third section discusses the similarities and differences among the variety of science teacher preparation programs and the future directions of science teacher preparation in Lebanon.

Education of Public School Teachers

Lebanese public school teachers are prepared in two institutions: The College of Education of the Lebanese University (hereafter referred to as College of Education) and CERD. The mandate of the College of Education is to prepare secondary school teachers while that of CERDs elementary and intermediate teachers' colleges is to prepare intermediate and elementary teachers. In what follows the programs offered by these two institutions will be described.

Programs Offered at the College of Education, Lebanese University

The College of Education offers the Certificat d'Aptitude Pédagogique à l'Enseignement Secondaire (CAPES) (Certificate of Qualification in Education for Secondary School Teaching) that is required for employment in public secondary schools. To be admitted to the CAPES program, students are required to hold a 4-year degree in a subject area taught at the secondary school level such as biology, chemistry, or physics[2] and to pass an entrance examination administered by the Council of Civil Service, a department of the Lebanese Government in charge of employment in the civil service.

As noted above, one of the requirements for admission to the CAPES program in the teaching of science is earning a 4-year degree in a science area. Many of the students who enter these programs graduate from the College of Science, Lebanese University or from other accredited universities. The program at the College of Science is divided into two 2-year cycles. During the first 2-year cycle, chemistry and biology majors take almost the same courses, while physics and mathematics majors take the same courses. Thereafter, students in each major take different courses, for the most part within their major discipline. Almost all courses taken during the four years required to complete the degree are in the major or in supporting areas.

To obtain the CAPES requires 74 semester credits in addition to a thesis (Table 3). The purpose of the thesis is to provide the student with the opportunity to investigate a science education topic in which he or she is interested (Faculté de Pédagogie, Université Libanaise, 1979). Writing the thesis, according to the professors who were interviewed, gives students the opportunity to integrate the different elements of the program. Thus a program in which education courses are taught independent of science content matter becomes much more coherent and meaningful.[3]

The general goals of the education programs offered at the College of Education include providing prospective teachers with theoretical and practical information needed for good teaching and helping them to develop skills necessary to live and work with others. Analysis of the course descriptions, the syllabi, and the interviews conducted with faculty members shows that almost all methods instructors emphasize the nature of science, constructivist ideas, the Lebanese Science Curriculum, and a variety of teaching and laboratory approaches to science teaching.

Most professors and department chairs interviewed asserted that a very strong subject matter background is extremely important for secondary

[2] Earth science is not taught as a separate subject in the Lebanese educational system. Starting at Grade 7, biology, chemistry, and physics are taught as independent courses.

[3] During the academic years 1995/96 and 1996/1997, the College of Education was charged with training a group of physics, chemistry, and biology inservice teachers. These teachers were contracted to teach at public schools during the last few years of the war in Lebanon and were not able to complete the CAPES. Because of the urgent need for these teachers, they were required to participate in a 24-credit accelerated training program which consisted of 16 credits of general pedagogy courses, four credits of methods, and four credits of field work.

Table 3. General Pedagogy Courses, Methods Courses, Field Work, and Subject Matter Requirements of Post Bachelor's Degrees at Six Lebanese Institutions

	Level	General Pedagogy	Science Methods*	Field Work	Subject Matter Requirements
American U. of Beirut	Secondary**	12 crs	6 crs	3 crs	BA/BS science
	Elementary	9 crs	3 crs	3 crs	24 cr. Science/math
Haigazian U.	Secondary & Elementary	12 crs	3 crs	3 crs	BA/BS
Lebanese American University	Secondary & Elementary	12 crs	3 crs	3 crs	BA/BS
Lebanese U.:					
--CAPES	Secondary	56 crs	6 crs	12 crs	4-year science degree
--DES	Secondary	36 crs	8 crs	--	4-year science degree
Notre Dame U.	Secondary & Elementary	12 crs	3 crs	3 crs	BA/BS science
University of Balamand	Secondary	18 crs	6 crs	6 crs	BA/BS science

*Methods courses usually include field components
**"Secondary" often includes preparation of Intermediate teachers

school teachers. In addition, they thought that more emphasis needs to be placed on introducing prospective teachers to effective teaching methods and laboratory activities related to science rather than teaching them general pedagogy. In the words of one physics education professor, "*Prospective teachers should first master the content of the subject they teach, then they should be introduced to effective science teaching methods. They need to use the lab as much as possible for applying what they learn*" (PHYS-LEB1[4]). In a similar manner, a chemistry education professor said,

> *Chemistry is an important component of the Lebanese curriculum; students need to understand all the concepts very well before they can teach these concepts. It is our responsibility to make sure they go to class well prepared in their subject matter. We also need to teach them as many methods as possible and make sure that they understand the reasons why these methods work.* (CHEM-LEB2)

[4] The following system was adopted to refer to the interviewees: the first part of reference gives the subject matter specialization, followed by the name of the institution and the interviewee number in that institution.

A biology education professor said: "*Students who come to us have a weak biology background. We need to help them remember their biology, understand it, and teach well to their students*" (BIO-LEB3). Finally, the Chair of the Education Department suggested that, "*Teaching is not only an art, it is a science. We need to use what research says about effective teaching and make sure that our pre-service teachers use the methods that work*" (CHAIR-LEB1).

The description of the CAPES program, as well as the opinions of the faculty members and department chair, suggests that the secondary science teaching program at the College of Education shares some of the characteristics of the academic and technological orientations because of the emphasis on preparing teachers who can convey subject matter efficiently using a variety of teaching techniques, methods, and strategies.

Programs Offered by the Center for Educational Research and Development (CERD)

CERD's mandate is to prepare elementary classroom teachers and intermediate science and mathematics public school teachers. To accomplish its mandate, CERD founded a number of elementary and intermediate teachers' colleges in different regions of Lebanon. The objective of the administrative unit in charge of these colleges is "to improve the knowledge, capability, and effectiveness of teachers . . . through inculcating new knowledge, educational doctrines, technology and methods" (Murr, 1983, p. 83). This indicates that the program has characteristics of both the academic and technological orientations.

Elementary teacher education programs[5]

There are two routes for preparing elementary school teachers by CERD, a 1-year program and a 3-year program, both of which prepare classroom teachers. The two programs have different admission requirements. While candidates for the 1-year program are required to hold the Lebanese Baccalaureate, those of the 3-year program are required to have completed Grade 9. The content matter preparation of the prospective teachers in the 1-year program takes place during their high school study. The one year spent at an elementary teachers' college is divided into two semesters. During the first semester, prospective teachers take courses in languages, social studies, mathematics, science, sociology of education, educational psychology, art, music, and physical education. In the second semester they take methods of

[5] The elementary and intermediate teacher preparation programs were frozen for a number of years between 1975 and the present because of the war in Lebanon.

teaching science, math, languages, and social studies in addition to practice teaching and field work.

Students in the 3-year program take all content courses at the elementary teachers' colleges. The program is divided into two cycles. The first cycle consists of two years and is dedicated to content preparation. The second cycle, which consists of one year, is identical to the 1-year program described above and is dedicated to professional education and field work.

Intermediate teacher education program

The intermediate teacher education program is a 3-year program beyond the Lebanese Baccalaureate, the first two years of which are devoted to content preparation in science while the third year is devoted to professional preparation. Content matter preparation takes place at the College of Science, Lebanese University, in which students follow the same curricula as those of majors in biology, chemistry, or physics. The third year is dedicated to professional education and is divided into three trimesters. During the first trimester students take courses in educational psychology, curriculum studies, laboratory, and methods of science teaching, in addition to school observation. The second trimester is devoted to field work and practice teaching. During the third trimester students return to the courses of the first trimester—educational psychology, curriculum studies, laboratory, science methods, and practicum.

Education of Private School Teachers

In the following sections, the programs offered in each university or college will be summarized, followed by a description of the curricula designed by the Ministry of Technical and Vocational Education. Note that the colleges and universities that offer 4-year programs have adopted, in most cases, a French model organized by years rather than by courses. Consequently, it is hard at times to translate requirements into course credits. For comparison purposes, percentages, and credit hours whenever possible, were used.

Programs offered at the American University of Beirut

The American University of Beirut (AUB) offers two programs leading to a Bachelor of Arts in Education/Elementary and Teaching Diplomas in elementary and secondary education.

Bachelor of Arts in Education/Elementary

The Bachelor of Arts in Education/Elementary at AUB requires a minimum of 96 credits to complete: 47% of course work is in education and related fields, 22% in cultural studies and languages, 25% in subject matter courses, and 6% in electives (Table 4). Education and related courses for the science/mathematics specialization are of two kinds: general pedagogy and specialized pedagogy. Courses in general pedagogy include sociology of education, psychology of education, philosophy of education, evaluation, and instructional procedures. Courses in specialized pedagogy include methods courses, observation, and field work. Science and science methods courses include a one three-credit science methods course, 18 credits in science and mathematics courses, and six credits in a practicum. Out of the 18 credits of science or mathematics, three credits have to be in each of mathematics and science and 12 credits can be selected from other science and mathematics courses. Consequently, a preservice teacher may take up to 15 credits of science in addition to an elective science course (Table 5).

Table 4. Percentage of Education, Cultural Studies, Subject Matter, and Electives in 3-Year Teacher Preparation Programs in Lebanon

	Education	Cultural Studies*	Subject Matter	Electives
American University of Beirut				
-- Elementary	47%	22 %	25 %	6%
Haigazian University				
-- Elementary	52%	29%	19%	--
-- Secondary	46 %	35%	19%	--
Lebanese American University				
-- Elementary	46%	22%	19%	13%
-- Science Education	20%	25%	43%	13%
Middle East College				
-- Elementary	36%	31%	14%	18%
-- Secondary	33%	31%	36%	--

*Includes other university requirements, such as language and religion

Teaching Diploma

The Teaching Diploma program is a 1-year post bachelor's program for elementary, intermediate and secondary school teachers. To obtain the Teaching Diploma/Elementary with a major in science/mathematics, preservice teachers are required to complete 24 credit hours in mathematics, science, or both mathematics and science, 9 credits in general pedagogy, 9 credits of methods courses, 3 credits of which are in science methods, and 3 credits of practicum (Table 3). Similarly, to obtain a Teaching Diploma/Secondary with a science major requires, in addition to a

Bachelor's Degree in Science, 12 credits in general pedagogy, 6 credits in science methods courses, and 3 credits in practicum (Table 3). The secondary methods courses are given for all science areas together. Students may opt to take a combined Bachelor/Teaching Diploma program during which they can take a number of the courses required for the bachelor's degree and the Teaching Diploma concurrently.

Table 5. Science Courses, Science Methods Courses, and Field Work Courses Required in 3 and 4-year Teacher Preparation Programs in Lebanon

	Level	Science	Science Methods	Field Work	Duration
American University of Beirut	Elementary	18 crs*	3 crs	6 crs	3 years
Haigazian University	Elementary	18 crs**	3 crs	3 crs	3 years
	Secondary	18 crs	3 crs	3 crs	3 years
Higher College for Teacher Preparation	Elementary	24 crs	4 crs	1.5 sems	4 years
Lebanese American University	Elementary	18 crs**	3 crs	3 crs	3 years
Lebanese University	(Grades 1-3)	3 crs	4.5 crs	24 crs	4 years
	(Grades 4-6)	6 crs	9 crs	13.5 crs	4 years
Middle East College	Elementary	15 crs**	3 crs	6 crs	3 years
	Secondary	39 crs**	3 crs	6 crs	3 years
University of Balamand	Elementary	6 crs***	3 crs	6 crs	4 years
University of Holy Spirit					
--Education license	Elementary	2 crs	2 crs	6 crs	4 years
--University license	Elementary	2 crs	2 crs	2 crs	3 years
University of Saint Joseph	Elementary	3 crs	3 crs	1 year	4 years

* crs = semester hour credits
** Can be science and/or mathematics
*** Not required, may be taken as electives

The specific goals of teacher preparation for teaching science at the elementary level, in the words of the elementary methods course instructor, include "*providing prospective teachers with an understanding of the goals of science education, the nature of science, science process skills, as well as teaching them methods and techniques specific to science with an emphasis on inquiry and hands-on/minds-on science activities*" (SC-AUB1). The Teaching Diploma/Secondary science program adopts a reflective practitioner approach to teaching and attempts to change teachers' conceptions about science teaching. In addition, preservice teachers are encouraged to plan and prepare lessons, practice, teach, and reflect on their teaching during classroom presentations as well as during field work. Moreover, according to the secondary science methods instructor,

> *Understanding science concepts is extremely important. This is why I emphasize curriculum-based science activities in the methods course. This will help prospective teachers understand the content and develop teaching skills at the same time. Prospective teachers need and want to learn techniques that can be used in the classroom. We help them get those techniques, but we also help them to think about the purposes of these techniques and how they help their students.* (SC-AUB2)

The science education program thus has many elements of the academic and technological orientations. However, the interviews show that it has some elements of reflective teaching and introduces students to a few constructivist ideas. The fact that reflective teaching is emphasized does not contradict the academic/technological nature of the program since "reflective teaching can be considered a generic professional disposition" (Feiman-Nemser, 1990, p. 221).

Programs offered at Haigazian University

Haigazian University (HU) offers two programs that lead to a Bachelor of Arts in Education/Elementary or Secondary and Normal Diplomas in elementary and secondary education.

Bachelor of Arts in Education/Elementary or Secondary

The Bachelor of Arts in Education/Elementary at HU requires a minimum of 94 credits to complete: 52% of course work in education and related fields, 29% in cultural studies, language, and other university requirements, and 19% in science courses. The science courses can be all in science or a combination of science and mathematics. The Bachelor of Arts in Education/Secondary requires a minimum of 94 credits: 46% in education and related fields, 35% in cultural studies, language and other university requirements, and 19% in science courses (Table 4). The science courses can be biology, chemistry, and/or physics.

Education and related courses required for the bachelor's degree include general and specialized pedagogy. Courses in general pedagogy include fundamentals of education, counseling psychology, history and philosophy of education, and introduction to educational administration. Courses in specialized pedagogy include methods courses in addition to observation and field work. Preservice teachers at the elementary level take one combined 3-credit course in principles and methods of teaching science, mathematics, and social studies in elementary schools with a 3-credit practicum.

Preservice teachers at the secondary level take one 3-credit methods course on teaching in secondary schools with a 3-credit practicum (Table 5). Elementary teachers graduating from this program typically teach math, science, and social studies while secondary level teachers teach either biology, chemistry, or physics and rarely a combination of two or three of these subjects.

Normal Diploma

The Normal Diploma program at HU is a 1-year post bachelor's program. To obtain the Normal Diploma requires, in addition to a bachelor's degree in a science area, 18 semester credits (Table 3). The science methods course at the elementary level is a combined science/mathematics course, while the course at the secondary level is a general secondary methods course.

The HU education program aims to provide the theoretical information and basic skills necessary for the beginning teacher. This is accomplished by providing students with the necessary background in general education, general and specialized pedagogy, subject matter, and practical and field work. Moreover, the program aims to satisfy students' curiosity, foster the development of critical thinking and open-mindedness, and prepare educational leaders (Haigazian University Catalogue, 1997-1999). The chair of the Education Program at HU indicated that their program has three foci.

> *The first and most important is the subject matter preparation, followed by the teaching techniques and the field work. We also emphasize critical thinking in all our programs. We help the prospective teachers understand the importance of using appropriate techniques in their teaching. But to do that they need a strong background in the subject they teach.* (CHAIR-HU1)

The program has some of the characteristics of both the academic and technological orientations, with an emphasis on critical thinking.

Programs Offered at the Higher College for Teacher Preparation (Makassed)

The Higher College for Teacher Preparation (HCTP) offers a 4-year program that prepares elementary classroom teachers and leads to an Education License in elementary education.

The program at HCTP requires attending approximately 40 periods per week for four academic years. Course work covers three major areas of study: (1) languages, social sciences, environmental education, health education, and cultural studies (58% of course work); (2) science (9% of course work), mathematics, and computers (9% of course work); and (3) education courses and related fields (23% of course work). In addition, preservice teachers spend almost 25% of their time in observation during the third year of the program and 50% of their time in practice teaching during the fourth year.

Education courses include general pedagogy and specialized pedagogy. Courses in general pedagogy include educational psychology and sociology, educational technology, and evaluation. Courses in specialized pedagogy include methods courses in all subject areas, observation, and field work. In science and science methods, preservice teachers take a total of 24 credits of biology, chemistry, and physics during the first and second years, and four credit hours of science methods during the third year (Table 5). The program consists of two 2-year phases. The first phase is completely dedicated to theoretical work, and the second phase includes an increasing amount of field and practical work. The HCTP teacher education program requires writing a thesis and participating in follow up activities during the first year of teaching. The thesis helps preservice teachers to investigate a topic of relevance to them, while follow-up activities during the first year of teaching are organized to give the new teachers an opportunity to identify and rectify problem areas in their teaching (Almakassed Benevolent Islamic Association [ABIA], 1997).

The goals of the program offered at HCTP are to prepare teachers who are first and foremost "educators who pay attention to students' psychological, social, and academic needs" (ABIA, 1997, p.1). This preparation requires teachers to be lifelong learners whose role is not to disseminate information but to plan, organize, and evaluate the educational process by involving students in their learning. The Director of Education at ABIA indicated that, "*Students who join HCTP need to have a strong content background. Also, they need to master the skills of effective teaching and demonstrate their ability to use these skills in real classrooms to succeed in the program*" (DIR- HCTP1). Because of these characteristics and the program description, the program at HCTP can be classified as having some characteristics of both the academic and technological orientations.

Programs Offered at the College of Education, Lebanese University

The College of Education of the Lebanese University offers a 4-year program that leads to a License en Sciences de L'Education (Education

License) for elementary private school teachers[6]. In addition, the College offers a Diplôme d'Etudes Supérieures (Diploma of Higher Studies) that prepares secondary private school teachers.

Education License

The College of Education offers the Education License in many specialties including Elementary Education--languages and other subjects (Grades 1-3) and Elementary Education--science and mathematics (Grades 4-6). Teachers at the Grades 1-3 level are prepared as classroom teachers, while those at the Grades 4-6 level specialize in different areas.

All programs leading to an Education License in elementary education require students to take 60 common credit hours in languages, general pedagogy, and general culture. Then, students majoring in a language and other subjects at the Grades 1-3 level take 60 more credits in specialized pedagogy, including methods courses in the area of their specialization and in field work. Similarly, students majoring in science and mathematics at the Grades 4-6 level take 60 more credits of specialized pedagogy (including methods courses), subject area, and field work. The 120 credit hours required for the Education License in Elementary Education are distributed over four years with the number of courses in specialized pedagogy and field work increasing as students advance in their studies.

The Grades 1-3 level teachers take courses in all subject areas including science, in addition to methods courses, field work, and other requirements. These students take three credit hours of science, four and one-half credit hours of science methods courses, and 24 hours of field work in all subject areas. The Grades 4-6 level teachers take 6 credits hours of science, 9 credits of science methods, and 13.5 credits of field work in science (Table 5).

The stated purpose of the program is to prepare teachers who are capable of helping their students to develop physically, psychologically, academically, and socially. This preparation requires that teachers acquire knowledge, skills, and attitudes necessary to develop personally and academically (College of Education, Lebanese University, 1997). Analysis of the science and science methods course descriptions and of the interviews conducted with science and science methods instructors shows that the purpose of teaching science and science methods is to help preservice teachers develop knowledge about the Lebanese Science Curriculum, lesson planning, goals of science teaching, and science teaching methods at the elementary level (See excerpts from interviews with professors at the

[6] The first year of the program described in this document was implemented in 1996/1997. The old program will be phased out gradually.

College of Education, Lebanese University, pp. 54-55). The program thus has some characteristics of academic and technological orientations.

Diploma in Higher Studies

The College of Education offers the Diplôme d'Etudes Supérieures (DES) (Diploma of Higher Studies). The Diploma is a 2-year post bachelor's degree open to prospective private school biology, chemistry or physics teachers at the secondary level. The DES requires a 4-year degree in a biology, chemistry, or physics.

To obtain the DES requires 44 semester credits and a thesis (Table 3). As in the case of CAPES, the purpose of the thesis is to investigates a science education topic of interest to the students (Faculté de Pédagogie, Université Libanaise, 1979). The methods courses are specific to the subject areas; consequently, there are biology, chemistry, and physics methods courses. The DES does not have designated credits for field work because most students enrolled in the program are practicing private school teachers.

Analysis of the course descriptions, the syllabi, and the interviews with science methods professors shows that almost all methods instructors introduce their students to the nature of science, constructivist ideas, the Lebanese science curriculum, and a variety of science teaching and laboratory approaches. In addition, the professors involved in the program suggest that the DES is more focused on research and research methodology. The program thus has some characteristics of the academic and technological orientations with emphasis on research activities (see interview excerpts pp. 54-55).

Programs offered at the Lebanese American University

The Lebanese American University (LAU) offers two programs that lead to a Bachelor of Arts in Education/Elementary and Teaching Diplomas in elementary and secondary education.

Bachelor of Arts in Education/Elementary and Science Education

The Bachelor of Arts in Education/Elementary at LAU requires a minimum of 92 credits to complete: 46% of course work in education and related fields, 22% in cultural studies and language, 19% in subject matter courses, and 13% in electives (Table 4). Education and related courses include general pedagogy and specialized pedagogy. Courses in general pedagogy include fundamentals of education; psychology of learning; testing, measurement, and evaluation; exceptional children; and guidance and counseling. Courses in specialized pedagogy include methods courses, observation, and field work. Science and science methods courses include one 3-credit course in combined science and mathematics methods, six 3-credit courses of science and mathematics courses, one of which is an elective science course, and three credits in a practicum (Table 5).

LAU offers a Bachelor's Degree in Science Education in two majors: general science (biology, chemistry, and physics) and biology/chemistry. This bachelor's degree requires a minimum of 92 credits, of which a minimum of 40 credits are in one science or a combination of sciences and 18 credits of education courses. The education courses are the same as those required for the Teaching Diploma (Table 4).

Teaching Diploma

The Teaching Diploma program at LAU is a 1-year post bachelor's degree. To obtain the Teaching Diploma in science requires, in addition to bachelor's degree in a science area, 18 semester credits of professional education (Table 3). The elementary methods course is a combined science/mathematics course, while the secondary course is a combined biology, chemistry, and physics methods course.

Education programs at LAU aim at preparing teachers who must "strive to have learners take an active role in their own education and inspire students to think, communicate, and participate in today's world" (LAU promotional brochure, no date). One science education faculty member indicated that he uses a variety of methods in his course to help students understand science concepts in general, and those concepts addressed in the curriculum more specifically, because many students do not have a good grasp of what they will teach. This faculty member also suggested that, "*Teaching science requires mastering a variety of techniques that will motivate students to learn science. They need to master those technique that have been shown to work well with students*" (SC-LAU1). Also, the Dean of the College of Arts and Sciences, which houses the teacher preparation program at LAU, said:

> *Our program at the secondary level requires students to have a BS in science before they join the program because we feel that students should be well informed in science. They have to understand science very well, before they take teaching courses that teach them how to teach. At the elementary level they need less science and math, but they have to know what they are going to teach. Understanding what is to be taught has the same, if not more, importance than how to teach. We help them learn and practice techniques that work with their students, but they have to know what to teach first.* (DEAN-LAU1)

According to the interviews conducted with the faculty members and the dean at LAU, the course descriptions, and program description presented in the catalogue, the science education program aims to provide preservice teachers with an adequate understanding of the nature of science and help them to encourage their students to think, construct knowledge, and find meaning in what they study. To accomplish the above, preservice teachers are introduced to teaching strategies aligned with the above goals such as inquiry teaching. The programs at LAU can thus be described as having characteristics of both the academic and technological orientations. However, some elements of constructivism permeate activities and education course offerings.

Programs Offered at Middle East College

Middle East College (MEC) offers two 3-year programs that lead to a Bachelor of Arts in Elementary Education and a Bachelor of Arts in Secondary Education.

The Bachelor of Arts in Education/Elementary with a major in science/mathematics at MEC requires a minimum of 105 credits: 36% of course work in education and related fields; 31% in cultural studies, language, and other college requirements; 14% in subject matter courses; and 18% in electives (Table 4). Science and science methods courses include one 3-credit science methods course, 15 credits of science and mathematics courses without specifying the number of credits required in each, and six credits in a practicum (Table 5).

The BA in secondary education requires 108 credits to complete: 33% of course work in education and related fields; 31% in cultural studies, language, and other college requirements; and 36% in subject matter courses (Table 4). The subject matter courses are the requirements of a major or two minors. Consequently, students can take a minimum of 39 credits in a science area or combination of sciences or 39 credits of mathematics and science (Table 5). This program requires one 3-credit science methods course and six credits of field work.

Education and related courses include general and specialized pedagogy. Courses in general pedagogy include philosophy of education, educational psychology, guidance and counseling, and evaluation. Courses in specialized pedagogy include methods courses in addition to observation and field work.

According to an interview with a MEC faculty member, the purpose of the education program at MEC is to provide teachers with the knowledge, skills, and attitudes required for teaching. The science methods course involves an in-depth study of structure, scope, methods, and materials for teaching science at the elementary or secondary levels (SC-MEC1). The program thus has the characteristics of academic/technological programs with some emphasis on attitude change.

Program Offered at Notre Dame University

Notre Dame University (NDU) offers one program that leads to a Teaching Diploma for elementary and secondary school teachers. To obtain the Teaching Diploma in science, students are required to complete, in addition to a bachelor's degree in a science, 18 credits in general pedagogy, methods courses, and a practicum (Table 3).

According to the Chair of the Education Department at NDU, the purpose of the Teaching Diploma program at NDU is to "*prepare teachers who can teach science for understanding, especially in an educational system that over-glorifies science and the memorization of scientific facts. This is accomplished by providing students with knowledge and skills necessary for good teaching and with field experiences that help teachers to be reflective and critical*" (CHAIR-NDU1). The program can thus be seen as having some characteristics of both the academic and technological orientations with emphasis on reflective teaching.

Program Offered at University of Saint Joseph

University of Saint Joseph (USJ), through its Institut Libanaise D'Educateurs (Lebanese Institute of Educators), offers a 4-year program, modeled after French programs, that leads to a License en Sciences de L'Education (Education License) for elementary teachers.

The program at USJ requires course work in three major areas of study (Farah-Sarkis, 1997): (1) cultural studies and other university requirements including computers (12% of course work); (2) subject matter of which 3% is in science (21% of course work); and (3) education and related areas (67% of course work). In addition, preservice teachers in the fourth year of the program participate in observation and practice teaching.

Education courses offered at USJ include general and specialized pedagogy. Courses in general pedagogy include educational psychology, developmental psychology, psychology of the child, and history and philosophy of education. Courses in specialized pedagogy include methods courses in all subject areas. In science and science methods, preservice teachers take three credits in science and three credits in methods of teaching science and/or mathematics (Table 5). In addition, students have to write a thesis, the purpose of which is to integrate the various components of the program. It is noteworthy that the catalogue of L'Institut Libanaise d'Educateurs does not list a science methods course in the program. However, the mathematics methods instructor indicated that preservice teachers are introduced to science methods either in the mathematics methods course or in the French language methods course (SC-USJ1).

According to the chair of the program, the education program at USJ emphasizes active involvement of preservice teachers in their learning in the hope that this orientation will translate into similar behavior in teaching young students. The course work, the field work, and the thesis are geared toward accomplishing this goal. The program offered by USJ is "*student-centered with emphasis on methods and techniques of teaching*" (CHAIR-USJ1); thus it can be classified as using a technological orientation with minimum emphasis on subject matter preparation.

Programs Offered at the University of Balamand

University of Balamand (UB) offers a 4-year program that leads to a License D'Enseignement en Sciences de L'Education (Education License) for elementary school teachers and a Teaching Diploma program for secondary school teachers.

Education License

The Education License at the UB requires 116 credits to complete: 62% of the course work in education and related fields; 30% in cultural studies, art, computers, languages, and health; and 8% credits in electives. In addition, students have to write a report on their field work equivalent to one credit hour. Preservice teachers may elect to take one or more science courses since they are required to take six credits of electives from courses outside the education field.

Education and related courses offered at UB include general and specialized pedagogy. Courses in general pedagogy are similar to courses offered at other universities. Courses in specialized pedagogy include methods courses in all subject areas, observation and field work. Preservice teachers at UB take three credit hours of science methods, six credits in a practicum, and have the option of taking up to six credits in science (Table 5).

Teaching Diploma

The Teaching Diploma program at UB is a 1-year post bachelor's degree program for intermediate and secondary school teachers. To obtain the Teaching Diploma requires, in addition to a bachelor's degree in a science area, 30 credits in general pedagogy, methods courses, and a practicum (Table 3). The methods courses are usually given for all science areas together. Students may opt to take a combined Bachelor/Teaching Diploma program during which they can take requirements of the BA and the Teaching Diploma concurrently.

One science education faculty member at UB stated that the goal of the Education License is "to prepare interdisciplinary teachers who are familiar with current pedagogical approaches and technical teaching skills" (SC-UB1). Also, the academic dean at UB stated that, "*The competencies evaluated in the program include students' subject matter knowledge, students' methodology of teaching, and relationships with students, parents, colleagues, and school administrators*" (DEAN-UB1). The program can thus be classified as academic/technological with more emphasis on the technological and on student-centered education. On the other hand, according to the same faculty member, the goal of the Teaching Diploma program at the UB is to prepare middle and secondary school teachers who have "*the knowledge and skills to teach effectively at this level*" (SC-UB1). Analysis of the course syllabi indicates that professors attempt to use a constructivist approach in their teaching. Moreover, there is an emphasis on involving cooperating teachers in evaluating preservice teachers field activities. The program can then be described as having some of the characteristics of both the academic and technological orientations.

Programs offered at the University of the Holy Spirit in Kaslik

The University of the Holy Spirit in Kaslik (USEK) offers a 4-year degree program, modeled after French universities, that leads to a License D'Enseignement (Education License) for elementary classroom school teachers. In addition, USEK offers a 3-year degree program that leads to a Licence d'Universite (University License).

Education License

The 4-year program at USEK requires attendance for 20 hours per week and is divided into two 2-year phases. During the first phase, preservice teachers take courses in philosophy, general sociology, general psychology, education, mathematics, religion, research methodology, computers, languages, and ecology. The second phase is dedicated almost completely to education courses and an increasing number of methods courses and field work.

Course work for the Education License at USEK covers three major areas: (1) languages, social sciences, and cultural studies (11.5 % of course work); (2) subject matter including science, mathematics, and computers (12.5% of course work); and (3) education courses and related fields (76% of course work).

Education courses offered at USEK include general and specialized pedagogy. Courses in general pedagogy include history of education, philosophy of education, educational psychology, educational laws, school sociology, educational economics, preschool education, general teaching strategies, planning in education, and educational research. Courses in specialized pedagogy include methods courses, observation, and field work. In science and science methods, preservice teachers take one 2-credit course in ecology and four credit hours of methods in all the subject areas, two credits of which may be dedicated to science and/or mathematics methods (Table 5).

University License

To obtain the University License in elementary education, students take the same courses as in the first three years of the Education License. Courses in the University License include: (1) languages, social sciences, and cultural studies (12% of course work); (2) science, mathematics and computers (17% of course work); and (3) education courses and related fields such as psychology, sociology, philosophy, and field work (71% of course work). The type of education courses required in the University License program are the same as those required in the Education License, with fewer education courses and less field work (Table 5).

According to interviews conducted with one USEK faculty member, education is a multi-disciplinary field to which philosophers, psychologists, sociologists, economists, business people, and education specialties have contributed. In addition, researchers in education have identified the best methods to teach and reduce educational problems (SC-USEK1). Therefore, it is no longer acceptable to reduce education to the transmission of information and to prepare one-discipline teachers. Instead, it is essential to prepare multi-disciplinary teachers who are well versed in different fields of knowledge that have contributed to education (SC-USEK1). The program offered at USEK is multi-disciplinary and student-centered with emphasis on methods and techniques of teaching. Thus it can be classified as using a technological orientation with minimum emphasis on subject matter preparation.

Curricula Designed by the Ministry of Technical and Vocational Education

There are two types of teacher education curricula designed and supervised by the Ministry of Technical and Vocational Education and implemented by private institutions: Technicien Superieur (Senior Technician) with a major in preschool and elementary education, and Baccalaureat Technique (Technical Baccalaureate). The Senior Technician program is a 3-year program beyond Grade 11, while the Technical Baccalaureate program is a 2-year program beyond Grade 11. Most teachers who graduate from either program work as preschool teachers. However, the Senior Technician degree prepares elementary school classroom teachers to teach at the Grades 1-3 level.

The present Senior Technician curriculum consists of academic and professional education courses distributed over a period of three years. The course work required for this program consists of 38% academic subjects and 62% professional education during the first and second years. During the third year, more emphasis is put on professional education, where it occupies approximately 79% of the course work. The Technical Baccalaureate program is similar to the first two years of the Senior Technician program.

There is no specific science methods course offered in either of the above programs. However, according to the director of an institution that offers these programs, students take methods of science teaching in the language methods courses or in courses designed to help teachers produce instructional aids. Moreover, science content may be included in health, environment, or nutrition courses.

Presently, the Ministry of Technical and Vocational education is in the midst of reform that aims to change the above programs. Since reform is still in its initial stages, a description of the new programs cannot be included in this document.

SIMILARITIES AND DIFFERENCES AMONG SCIENCE TEACHER PREPARATION PROGRAMS

To obtain the Teaching Diploma, Normal Diploma, Certificat d'Aptitude Pédagogique à l'Enseignement Secondaire (CAPES) (Certificate of Qualification in Education for Secondary School Teaching), or Diplôme d'Etudes Supérieures (DES) (Diploma of Higher Studies), prospective teachers are required to have an undergraduate degree in a science area or a number of science courses depending upon the classes they intend to teach. Furthermore, they have to complete general pedagogy courses, science methods courses, and field work.

These programs are similar in that they emphasize subject matter preparation and introduce students to diverse teaching techniques and strategies in the methods courses. Likewise, they are similar in their lack of emphasis on field work. Emphasis on subject matter preparation is due to the fact that most post bachelor's programs prepare secondary school teachers.

The differences among the post bachelor's programs are in the number of credits in each of the courses and in the thesis requirement. While the degrees offered at the Lebanese University require a thesis and a 4-year science degree beyond the Lebanese Baccalaureate, those offered at the other universities and colleges require a 3-year degree beyond the Baccalaureate and no thesis. Other differences include the number of credits of methods and field work, which ranges between three and six credits for the methods courses and 3 and 12 credits for field work. The amount of science required in the elementary and secondary programs differs among the programs: while most secondary programs require an undergraduate degree in a science area or a combination of sciences, the elementary level programs require less science and prepare science and mathematics teachers rather than just science teachers (Table 3).

Most of the 3- and 4-year programs offered by universities and colleges in Lebanon prepare elementary school teachers. These programs are similar in that they require at least one science methods course equivalent to two or three credits -- except the program at the Grades 4-6 level offered by the Lebanese University, which requires nine credit hours. Also, all programs require field work. However, the amount of time devoted to field work varies appreciably between one program and another. For example, there are programs that require one full year, others that require almost 1.5 semesters, and still others that require only two credit hours. The number of required science courses varies among universities: universities that follow the American model of higher education require more science than those following the French model, with one exception (Higher College for Teacher Preparation). Furthermore, secondary science education programs require significantly more science than elementary ones (Table 5).

Although the intermediate level of schooling is defined by Lebanese law, programs aimed specifically at intermediate teacher preparation are scarce. This may be due to the perception of some educators that the needs of intermediate students are similar to those of secondary, hence programs such as AUB. Others believe that the special needs of intermediate students are met by special programs, such as those in the CERDs. Other reasons may be financial: private universities cannot afford to offer differentiated programs for intermediate level students.

Generally, 3-year programs are offered at universities which follow an American model of higher education. These programs are 3-year programs because the Lebanese Baccalaureate is equivalent to the freshman class and students holding the Baccalaureate are admitted to the sophomore class. In addition, these 3-year programs require more subject matter than 4-year programs offered at universities which follow a French model. It appears that universities which follow a French model emphasize education, psychology, and sociology significantly more than those which follow an American model. One possible reason for this difference is that many teachers who graduate from universities which follow a French model teach languages rather than other subjects, while the teaching of science and mathematics is relegated to specialists in science or mathematics education, or teachers who hold science and mathematics degrees.

The Lebanese University stands out among other universities and colleges in that it offers two types of 4-year programs at the elementary level: one for Grades 1-3 and another for Grades 4-6. This structure is a direct response to the new Lebanese Educational Ladder, which has two stages at the elementary level. Consequently, because the second stage (Grades 4-6) requires a significant amount of science teaching, the Lebanese University has adopted a program that meets the needs of teachers at this level. Also, the CAPES and the DES programs offered at the Lebanese University have the distinct characteristic of requiring a 4-year science degree, a significant amount of general pedagogy, and a thesis, which make them very demanding programs. What is evident is the limited importance placed on field work in these two programs.

Another program that stands out is the one offered by the Higher College for Teacher Preparation, which incorporates a significant amount of science together with 1.5 semesters of field work, a thesis, and a 1-year induction program (Table 5). Finally, The Lebanese American University offers a unique program, the BA in Science Education, that may be a good model for the preparation of intermediate school teachers (Table 4). This program combines a significant amount of general science content with education courses and may be useful for biology, chemistry, and physics teachers at the intermediate level.

In summary, science teacher preparation programs in Lebanon are characterized by the following:

1. Post-graduate programs that prepare secondary teachers with significant amount of science background, especially those programs offered at the Lebanese University.
2. Three- and four-year programs that prepare classroom teachers or science/mathematics teachers for all elementary grades, except the programs offered at the Lebanese University
3. Absence of university level programs for the preparation of intermediate school science teachers.
4. Requirement of a thesis in many 4-year programs and in the post bachelor's programs offered by the Lebanese University.
5. Lack of emphasis on field work.
6. Adoption of an orientation which has some characteristic of the academic and technological orientations to teacher preparation, without neglecting emphasis on constructivism, reflective practice, thinking, and inquiry.

FUTURE DIRECTIONS FOR SCIENCE TEACHER PREPARATION

What are the future directions of science teacher preparation programs in Lebanon and what are the prospects for improving these programs? The present curriculum reform movement in all subject areas, spearheaded by the Center for Educational Research and Development, has the potential to spur significant changes in Lebanese teacher preparation programs. In response to the new science curricula, private and public institutions are bound to change the structure and content of teacher preparation programs. For example, institutions may have to prepare specialized teachers for each of the stages of the new educational ladder, because students at each of these stages have different academic, social, and psychological needs. One outcome should be designing special programs for the preparation of lower elementary, upper elementary, and intermediate teachers, because students at each of these levels have specific academic, social, and psychological needs that are different from each other and from secondary students. Another outcome should be requiring certification for teaching in private school. Presently private school teaching does not require certification. Consequently, many private schools employ beginning biology, chemistry, or physics teachers to teach science at all levels, resulting in an over-emphasis on disseminating information (because teachers tend to teach the same way they were taught). Finally, new teacher preparation programs need to put more emphasis on field work and collaboration with schools in the preparation of teachers. This will require new structures and new ways of thinking about the roles of both universities and schools in teacher preparation and in the professional development of teachers.

REFERENCES

American University of Beirut. (1996-1997). [Catalogue]. Beirut, Lebanon: Author.

Almakassed Benevolent Islamic Association. (1997). *History of the founding of the Higher College for Teacher Preparation*. Beirut, Lebanon: Author.

Anderson, R., & Mitchener, C. (1994). Research on science teacher education. In D. Gabel (Ed.), *Handbook of research on science teaching and learning* (pp. 3-44). New York: Macmillan Publishing Company.

Notre Dame University. (1996-1997). [Catalogue]. Louaize, Lebanon: Author.

Lebanese American University. (1996-1997). [Catalogue]. Beirut, Lebanon: Author.

Center for Educational Research and Development. (1995). *Al-Haykalia Ajadida Lita'alim Fi Lubnan* (The New Lebanese Educational Ladder) . Beirut, Lebanon: Author.

Eido, R. (1983). I'idad mualimi alulum wariyadiyat fi Jami'iyat Al-Makassed Alkhairiyyah fi Beirut (Science and mathematics teacher preparation at the Almakassed Benevolent Islamic Association in Beirut). In M. Jurdak (Ed.), *Mathematics and science education in the elementary and intermediate cycles in Lebanon* (pp. 88-91). Beirut: Science and Mathematics Education Center, American University of Beirut.

Faculté de Pédagogie, Université Libanaise [College of Education, Lebanese University] (1979). *Diplome délivrés par la faculté* [Degrees Offered by the College of Education] [Brochure]. Beirut: Lebanon: Lebanese University.

Farah-Sarkis, F. (1997, May). *Dawr Ata'alim fi Tatwir Anazzam Atarbawi: Halat Lubnan* [The role of higher education in developing education systems: The case of Lebanon]. Paper presented at the Regional Arab Symposium on the Role of Higher Education in Developing Education Systems, Beirut, Lebanon.

Feiman-Nemser, S. (1990). Teacher preparation: Structural and conceptual alternatives. In W. Houston, M. Haberman, & J. Sikula (Eds.), *Handbook of research on teacher education* (pp. 212-233). New York: Macmillan Publishing Company.

Freiha, N. (1997). Mukaranat Manahij atarbiyah [Comparison of Education Curricula]. In A. El-Amine (Ed.), *Atta'alim Ala'ali fi Lubnan* [Higher Education in Lebanon] (pp. 273-295). Beirut, Lebanon: Lebanese Association for Educational Sciences.

Haddad, M. (1983). The preparation of mathematics and science teachers at the American University of Beirut. In M. Jurdak (Ed.), *Mathematics and science education in the elementary and intermediate cycles in Lebanon* (pp. 92-93). Beirut: Science and Mathematics Education Center, American University of Beirut.

Haigazian University. (1997-1999). [Catalogue]. Beirut, Lebanon: Author.

Kennedy, M. (1990). Choosing a goal for professional education. In W. Houston, M. Haberman, & J. Sikula (Eds.), *Handbook of research on teacher education* (pp. 813-825). New York: Macmillan.

Kuliyat Attarbiya, Al-Jamia'a Alubnaniyya [College of Education, Lebanese University]. (1997). *Manahij Al-Ijazza Atta'alimiya Fi Attarbiyya* [Curricula of the Education License]. Beirut, Lebanon: College of Education.

L'Annuaire de L'Université Saint-Esprit de Kaslik [Catalogue of University of the Holy Spirit in Kaslik]. (1996-1997). Kaslik, Lebanon: USEK.

L'Annuaire de L'Université Saint-Joseph [Catalogue of University of Saint Joseph]. (1996-1997). Beirut, Lebanon: USJ.

Middle East College Bulletin. (1996-1998). Sabtieh, Jdeidet El-Matn, Lebanon: MEC.

Murr, G. (1983). The preparation of mathematics and science teachers for public schools. In M. Jurdak (Ed.), *Mathematics and science education in the elementary and intermediate cycles in Lebanon* (pp. 82-87). Beirut: Science and Mathematics Education Center, American University of Beirut.

Tahdid Manahij Ata'alim Ala'am Ma Kabl Ajami'i Wahadafiha [Curricula and goals of pre-college education]. (1997). Lebanese Public Law No. [illegible].

Tanzim Kuliyyat Attarbiya Filjamia'a Lubnaniyyah [Organization of the College of Education in the Lebanese University]. (1979). Lebanese Public Law No. 1833.

University of Balamand. (1997). [Catalogue] Balamand, Lebanon: Author.

Wehbe, N. (1984). Al-i'idad atarbawi fi Lubnan wali'idadad fi kuliat atarbiya [Educational training in Lebanon and training in the College of Education]. *Recherche Pédagoguique, 12*, 73-100.

Chapter 5

Science Teacher Education in Pakistan

Policies and Practices

Hafiz Muhammad Iqbal and Nasir Mahmood
University of the Punjab-Lahore, Pakistan

Abstract: The chapter addresses the problems faced by science teacher educators in Pakistan. We first discuss the historical background of science education development to give readers an understanding of the Pakistani perspective. We argue that teacher education programs are caught up in a vicious cycle. On the one hand, increasing the number of qualified teachers is a natural need for the expanding system of education, and, on the other hand, there is a genuine concern for the maintenance of quality in the education imparted. Obviously, if teacher education programs are not carried out with care, quality aspects will be compromised as in the past. This requires an overall improvement in (a) the quality of course content, (b) the minimum duration, and (c) the admission requirements for different programs of teacher education. We discuss the disparity between pronounced policies of teacher education and their actual implementation and suggest methods to improve science teacher education.

Science teacher education in Pakistan is confronted with many problems. Some problems are common to many other countries in the world, while some are unique to a developing country like Pakistan, whose education system has not yet been established on realistic footings. These problems can be fully understood in the historical and social perspectives of the education system in general. Pakistan as an independent state came into being just half a century ago. The nation celebrated its 50th anniversary on August 14, 1997.

Thus the education system of the country does not have a long history, although it is older than the nation itself. The British rulers introduced this education system during the second half of the nineteenth century when Pakistan was still part of the Indo-Pak sub-continent. This does not mean that there was no education system in the sub-continent prior to the establishment of the English education system. The fact is that the

S.K. Abell (ed.), Science Teacher Education, 75–92.

indigenous system of education was replaced by the English system of education. The British system was installed to achieve colonial purposes. This system was designed to produce a literate manpower to assist the colonial rulers at the lower levels of governmental and economic administration. Higher level jobs were reserved for English men only. As Hayes (1987) pointed out, such an education system was meant only for the privileged few, who were supposed to govern the masses rather than to serve them. Because the real intention was to produce white-collar workers, undue emphasis was placed on liberal arts and non-professional education. This system served the narrow utilitarian purpose of the colonial rulers very well, but had no room for nourishment of the individuality, creativity, or intellectual capacities of learners. The same was reflected in the curricula and textbooks, which were rigidly oriented towards memorization and passing examinations.

Science as a school subject was not introduced in the British system of education in Pakistan at the beginning, although it had been fighting for its recognition for quite a considerable time. Later, science subjects were introduced from top-down, first at the higher level and then at the lower level. The science taught at the secondary level was didactic and theoretical in nature. As far as the primary and middle school levels were concerned, practically no science was taught until the 1950s. The main emphasis at this level of education was upon reading, writing, and arithmetic--the traditional three R's (Government of Pakistan, 1975). Because science education had very little or no emphasis in the curriculum, there was no clear policy or system of education for teachers in science.

Soon after independence, it was realized that Pakistan's education system was not based on realistic objectives. The system, with an emphasis on the three R's and liberal arts, was more geared to serving colonial purposes. In order to serve the purposes of an independent state, the education system needed an overhauling and restructuring, with a greater emphasis on science and technology. It was also realized that the curricula at various level of education lamentably ignored science, technical, and vocational subjects. The first Pakistan Education Conference, held in 1947 in Karachi, prepared the ground for major changes in the education system of the country. The conference received a message from the founder and first Governor General of Pakistan, Muhammad Ali Jinnah, emphasizing the need for giving the education system of the country a scientific and technical base in order to build up the economic life of the newly liberated state. It was believed that a strong science and technological education was imperative for achieving economic development and prosperity. During the same conference, a Committee on Scientific Research and Technical Education was convened. The committee discussed the problems and issues of science and technical education and recommended that every effort be made to promote fundamental as well as industrial research (Government of Pakistan, 1947). Ironically, the agenda before the committee was to discuss problems and issues of science, technology, and research at the higher level. These deliberations were not related to science education, particularly at the school level, nor was the issue of science teacher preparation placed on the agenda of the conference.

The second effort in this regard was the setting up of a commission in 1959 that was assigned the job of recommending measures to improve the system of education in Pakistan. Observing the condition and status of science education in the country, the commission recommended that, "In an age when science and applied technology determine the rate of progress of the nations, the teaching of science and mathematics be given a strong base in our schools" (Government of Pakistan, 1959, p. 122). The commission recommended that science and mathematics education be made compulsory for grades 6-10. The commission also noted an imbalance between different components of the curriculum at the primary level. The commission recommended that an equal emphasis be given to science, mathematics, and the liberal arts at the primary level. Following the recommendations by the commission, nature study was introduced at the elementary level to familiarize students with the environment by direct observation. The component of science, however, remained very weak, and in actual classroom practice the major emphasis still remained on the three R's. Theoretically, in the late fifties and early sixties, science education was made a compulsory part of the curriculum for grades 1-8. However, the quality of instruction in science at all levels remained far from satisfactory. The reason for this poor quality of instruction can be traced to the lack of proper education of science teachers.

The first policy that focused science education as discipline of human inquiry was the National Education Policy 1979 (Government of Pakistan, 1979). This policy discussed science education as a separate component of secondary education and emphasized its development on sound footing. Regarding the quality of science instruction in the country, the policy observed:

> In spite of several curricular reforms in science education, the quality of instruction in science education, particularly at the pre-university levels, has not improved considerably. This is so because science is still being taught as a "dogma". Very little curiosity in scientific inquiry, initiative and involvement in understanding the scientific concepts and processes are emphasized. (p.35)

This policy was different than other previously announced policies in the sense that, for the first time, the quality of pre-university science education was discussed and measures were recommended to improve it. In previous policies the main emphasis had been on the development and improvement of science and technology education at the university level. This was based on the assumption that to achieve economic progress, it was more important to give emphasis to secondary and higher education than to primary and middle level education. Measures adopted to improve school science education and their impact will be discussed subsequently.

THE CURRENT STATUS OF SCIENCE EDUCATION IN PAKISTANI SCHOOLS

The structure of the Pakistani education system is depicted in Figure 1. At present, science education in Pakistan is compulsory at elementary (primary and middle) and optional at the secondary levels. From grades 1-8, science is supposed to be taught in an integrated manner, comprising some components from biology, chemistry, physics, and Earth science. At the primary level, science should occupy about 12% of the total school time. At the middle school level, 13-15% of instructional time is allocated to science. At the secondary level, about 12-14% of the time is allotted to each science subject-- physics, chemistry, and biology--with a total time for science being about 40%. The purpose of science education at the elementary level is to familiarize students with nature and their environment. At the secondary levels, grades 9 and 10, science is taught in the form of separate subjects--physics, chemistry, and biology. The main purpose of secondary and higher secondary levels of education is to prepare students for tertiary education in different scientific disciplines, largely in the engineering and medical fields. Very few students take science subjects with a particular interest and purpose of following pure science at the higher level. A large teaching force has been required to teach science, not only at the primary and middle levels where science is a compulsory component, but at the secondary level as well, where it is optional. Such a teaching force is not available even at this time, nor do teachers who have been entrusted with the job of science teaching have adequate preparation to meet the demands of teaching science in the changing world. A cursory look on the teacher education program in Pakistan will reveal this fact.

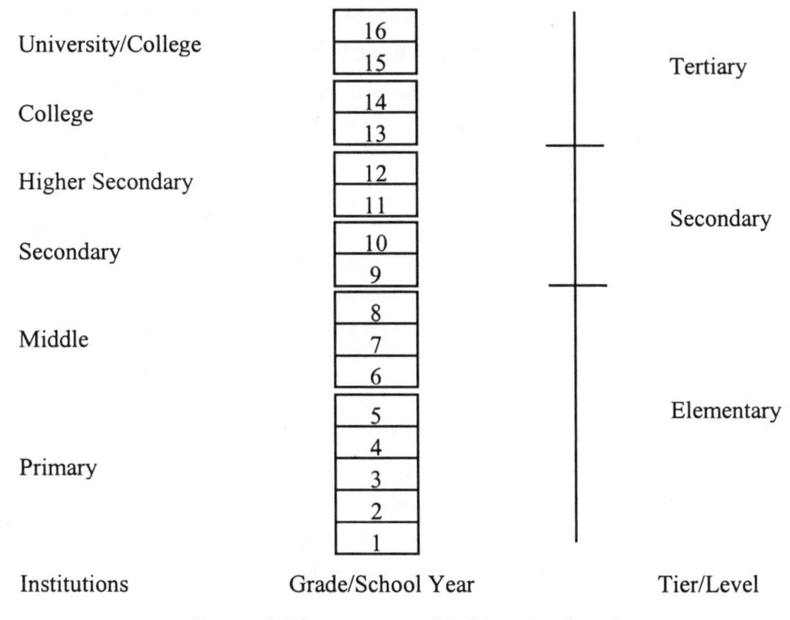

Figure 1. The structure of Pakistan's education system

Preparation and Qualification of Teachers

After independence, and as a result of recommendations made by various commissions, a rapid expansion in the education system in Pakistan was made. This speedy growth of the education system has been accompanied by important advances in teacher education. The general strategy in teacher preparation programs has been directed by two requirements: (a) increase the number of teachers, especially at the primary level, and (b) improve the quality of teacher preparation. On both accounts substantial progress has been made. Yet, when viewed from the overall requirements of teachers at various educational levels, teacher preparation programs have suffered from stagnation and slow growth. The Education Conference of 1947 emphasized the need for a properly educated and reasonably well paid teaching force, and recommended that provinces should take necessary steps in this regard. Similar emphasis has been placed on teacher education in all subsequent policies and plans. The report of the National Commission on Education (Government of Pakistan, 1959), in particular, identified the gaps in teacher preparation programs, and called for a massive increase in such programs. Regarding the preparation of teachers, the commission was of the view that the duration of preservice education of teachers at the primary level be extended to two years, instead of one year. However, the commission resolved that, in order to meet the shortage of teachers, the practice of one year preparation be continued for a period of five years. For teachers of elementary classes, the commission recommended that they be given two years of preparation. Similarly, it was recommended that the B.Ed. program for teachers also be extended to two years. The commission did not discuss the preparation of science teachers separately.

The Education Policy 1972 (Government of Pakistan, 1972) envisaged a massive shift from general to scientific and technical education both at the secondary and higher secondary levels. The rapid expansion of education at all levels and the shift to science education necessitated expansion in teacher preparation programs. Because of the lack of long-term planning and the tendency of the successive governments to adopt short-term measures, it was not anticipated that a large teaching force would be needed. Thus, instead of expanding the opportunities of preservice preparation for teachers, a short-term measure was adopted. It was recommended to recruit low-qualified teachers and provide them with an intensive inservice education to make up their deficiency in academic knowledge and pedagogic skills. Despite the emphasis of the policy on the shift from general to scientific and technological education, the policy did not address the problems of science education at the school level. Nor did the problems of inadequacy in science teacher programs get prominence in the recommendations made in the policy.

The Education Policy of 1979 (Government of Pakistan, 1979), on the other hand, declared that no system of education could be better than its teachers. The quality of teacher education was considered of utmost importance. Improvement was sought in teacher preparation programs in science, mathematics, industrial arts, agriculture, commerce, and home economics. New programs were proposed to strengthen both preservice and inservice teacher education. A national coordination committee for inservice education was also proposed in order to suggest measures to improve the inservice education offered to teachers at various levels.

In 1992, the government announced another National Education Policy. The policy, like many similar policies of the past, stated its dissatisfaction with the quality of teaching in schools and recommended improvement through the professional growth of teachers. This policy did not address the issue of science teaching at the pre-college level separately. However, it mentioned the teaching of science at the tertiary levels. Regarding teacher education, the policy made three particular recommendations. Firstly, it announced that mobile training units would be set up to provide training to inservice teachers. Secondly, the policy pronounced that teacher education curricula would be revised and the need for increasing its duration would be assessed. Thirdly, it declared that the teaching practice of teachers would be further strengthened through a regular inservice training adopted for teachers at all levels (Government of Pakistan, 1992). Some of these recommendations are in the implementation phase and some are yet to be implemented.

Preparation of Elementary School Teachers

The origin of teacher preparation programs goes back to the establishment of Normal Schools in the latter half of the 19th century. At present, elementary level teacher preparation programs are being offered in 87 institutions. There are two kinds of programs: the PTC (Primary Teaching Certificate), and the CT (Certificate of Teaching). The PTC program is meant for the teachers who teach in grades 1-5. The duration of this program is of one academic year (48 weeks). Admission to the PTC program requires that the applicant should be matriculate (10 years schooling). The CT program prepares teachers to teach all subjects up to 8th grade, including English. It is also of one academic year duration. The condition for admission to the CT program is an F.A. (Fellow of Arts) or F.Sc. (Fellow of Science) certificate, that is, 12 years of schooling (Government of Pakistan, 1977).

Preparation of Secondary School Teachers

The institutions preparing secondary school teachers are known as Colleges of Education, and those offering advanced education leading to the M.A. Education and M.Ed. degrees are called Institutes of Education and Research, or Departments of Education. These are usually affiliated with universities. At the time of independence, there was only one college of education for men and one for women in the area now called Pakistan. There was no Department of Education or Institute offering master's degrees in education. The first such institute, the Institute of Education and Research (IER) was established in 1960 at the Punjab University, Lahore in collaboration with Indiana University (USA), following the recommendation by the National Commission on Education (Government of Pakistan, 1959). Other IERs or Departments of Education were subsequently established. At present there are four Institutes of Education and Research, five Departments of Education, and 11 Colleges of Education preparing secondary school teachers (Government of Pakistan, 1989). Two types of programs are offered for the preparation of secondary school teachers, requiring differences in the years of general education and professional training: (a) 1-year B.Ed. (Bachelor of Education) program (14+1 model); and (b) 3-year B.S.Ed. (Bachelor of Science Education) program (12+3 model). For the 1-

year program, the minimum qualification required for admission is a B.A./B.Sc. degree. In the 12+3 model, the minimum qualification required for admission is the F.Sc. This model has been introduced in two colleges and is meant for science teachers only. These programs were approved by the National Committee on Teacher Education and were implemented in 1976-77. In spite of all these efforts to revise the courses of studies in teacher education, the programs have invited a lot of criticism. The courses which prospective teachers undergo are deficient both in respect to content and duration. The academic and professional qualifications required for the recruitment of teachers at various levels are shown in Table 1.

Table 1. Academic and Professional Qualification of Teachers at Different Levels

Level of teaching	Type of Examination	Academic Qualification	Professional Preparation
Primary	5th Class	Matric. (secondary)	PTC
Middle	8th Class	F.A./F.Sc.(higher sec.)	CT
Secondary (High)	Matric	BA/B.Sc.	B.Ed./B.S.Ed
Intermediate	F.A./F.Sc	MA/M.Sc.	---
Degree Colleges	BA/B.Sc.	MA/M.Sc.	---
University	MA/M.Sc.	MA/M.Sc./Ph.D.	---

SCIENCE TEACHER EDUCATION IN PAKISTAN

Before independence, because science was not a part of the school curriculum, teacher preparation institutions did not have any separate program for science teachers. Even after independence, this policy continued for a considerable period of time. As Table 1 depicts, the current academic qualification required to teach at primary level is a secondary school certificate. Prior to the 1950s, science was not taught in primary and middle schools, while at the secondary levels, science was an optional subject. Thus many of the teachers recruited to teach at the primary and middle levels did not have any science background. Thus, when science was introduced at the primary and elementary levels, its teaching was entrusted to non-science graduates. The situation has become worse recently. With the rapid increase in the number of schools, and the introduction of science at all levels of school education, the number of science teachers required has expanded. At the primary and elementary levels, no separate science teacher is envisaged. At the middle level, one science teacher per school is an accepted norm. However, at the secondary level, the situation becomes more complex, because science is taught as independent subjects of biology, chemistry, physics, and mathematics. The ideal situation would require four teachers per school, one for each subject. This ideal has not been attained, nor is it likely to be attained in the near future. Two science teachers per school is the existing norm. But, as shown in Table 2, a substantial number of science teachers both at the elementary and secondary levels remain untrained.

Table 2. Ratio of Trained and Untrained Science Teachers in Elementary and Secondary Schools in Pakistan in Three Different Years

Year – Level	Number of Schools	Number of Teachers		
		Trained	Untrained	Total
1983-84				
Secondary	4,000	6,000	3,000	9,000
Elementary	6,000	600	5,400	6,000
Total	10,000	6,600	8,400	15,000
1987-88				
Secondary	5,000	6,500	3,500	10,000
Elementary	6,500	1,000	5,500	6,500
Total	11,500	7,500	9,000	16,500
1992*				
Secondary	5,500	7,000	4,000	11,000
Elementary	7,000	1,500	5,500	7,000
Total	12,500	8,500	9,500	18,000

*Projected figures — Source: Ministry of Education 1989

In the past, all secondary school teachers were recruited in the same cadre or pay scales, by and large on their merit (i.e., academic achievement, irrespective of qualification in science or arts). Arts graduates were available in greater number than science graduates, partly because of low enrollment and high dropout in science, and partly because science graduates aspired to more promising jobs than the teaching profession. Since the early 1980s, the situation has slightly improved, but not to a satisfactory level. Even at present, many of the teachers actually engaged in science teaching are not themselves qualified in science and/or do not possess professional qualification. Commenting on the situation, UNESCO (1984) reiterated that a good proportion of science teachers did not possess degrees in science. In some cases, teachers are teaching science subjects which they themselves have never studied as students. The problem becomes more acute in schools situated in rural areas, where the majority of teachers teaching science, particularly at the secondary level, have little or no background in science. For example, in the Punjab province in 1980, there were a total of 15,969 graduate teachers in 1,844 high schools. Out of them only 1,444 were science graduates. The percentage of science students enrolled in that year was 46.3 (National Education Council, 1987). This means that in terms of workload, 9% of trained science graduates were sharing 52% of the entire workload.

Table 3 reflects the situation in 1986 in the Punjab, a highly populated province of the country. The table shows that a total of 325 professionally prepared teachers were available to teach science in 3,116 elementary schools, and 2,611 professionally prepared teachers were available to teach science in 2,020 secondary schools in 1986. The table also shows that a large number of teachers employed in elementary/secondary schools was not professionally prepared for their assignment. Though there was a shortage of professionally qualified teachers in all schools, the situation was more critical in elementary schools, where science education is a compulsory component of the curriculum. The ratio of science vs. arts teachers is not sufficient to share the teaching load of about 47% science students. This results in assigning the teaching of science classes to non-science teachers. These figures pertain to Punjab, the largest province of Pakistan. The situation in the rest of the country is worse, except in the urban Sindh, particularly the city of Karachi.

Table 3. Number of Science Teachers in Elementary and High Schools of the Punjab in 1986

	Elementary			Secondary		
	Rural	Urban	Total	Rural	Urban	Total
Schools	2783	333	3116	1185	835	2020
Male Teachers	1887	131	2018	883	479	1362
Female Teachers	896	202	1098	302	356	658
Science Teachers	4524	1344	5868	1906	2221	4127
Male	3104	615	3719	1470	1390	2860
Female	1420	729	2149	436	831	1267
Trained Science Teachers	110	215	325	724	1887	2611
Male	100	174	274	673	1390	2063
Female	10	41	51	51	497	548

Source: Pakistan Science Education Project for Secondary Education (1986)

SCIENCE INSTRUCTION

The second problem noted in government reports was the low quality of instruction in science. This issue was observed by the Education Policy 1979. The policy stated that the low quality of instruction in science was due to the lack of competence of science teachers. The policy document reiterated:

> Many teachers lack desired knowledge, competencies, skills and scientific attitudes. As such teacher's demonstration and inquiry-directed experiments seldom find their way into classrooms and laboratories. (Government of Pakistan, 1979, p. 35)

The policy document stressed constant maintenance of good instruction, particularly at the pre-university level. To improve the quality of science education in Pakistan, two important measures were taken. Firstly, the recommendation made by the National Committee on Teacher Education in early 1974 to start a 12+3 (twelve years of general education plus three years of professional training) model of science teacher education was implemented. However, this model has been introduced in only two Colleges of Education, one in the Punjab province and second in the Federal Capital Islamabad. The number of students graduating from these colleges are far fewer than the requirements of the system. However, along with the introduction of the 12+3 model of science teacher education, another important breakthrough was the introduction of the master's degree program in science education at the Institute of Education and Research, University of the Punjab, Lahore. Two degree programs were introduced in 1987, a 1-year M.Ed. (Science) degree program for inservice teachers holding a bachelor's degree in science education (B.Ed. or B.S.Ed.). The second is a 2-year program of M.S.Ed. for students having a B.Sc. in natural sciences. The present structure of teacher education in general and for science teachers as well is as depicted in Figure 2.

The second reform instituted under the 1979 policy was to institutionalize the efforts for development of science education in Pakistan. It was proposed that a National Center for Science Education be established. This center was to be given the responsibility of conducting research in science education catering to the development of science education professionals, and supervising science education development efforts in the country. Subsequently, the Science Education Project was launched in the mid 1980s with a loan from the Asian Development Bank. Under this project, new science curricula were developed for grades 6-10. An Institute for Promotion of Science Education and Training (IPSET) was also established to deliver inservice education to science teachers and coordinate the efforts

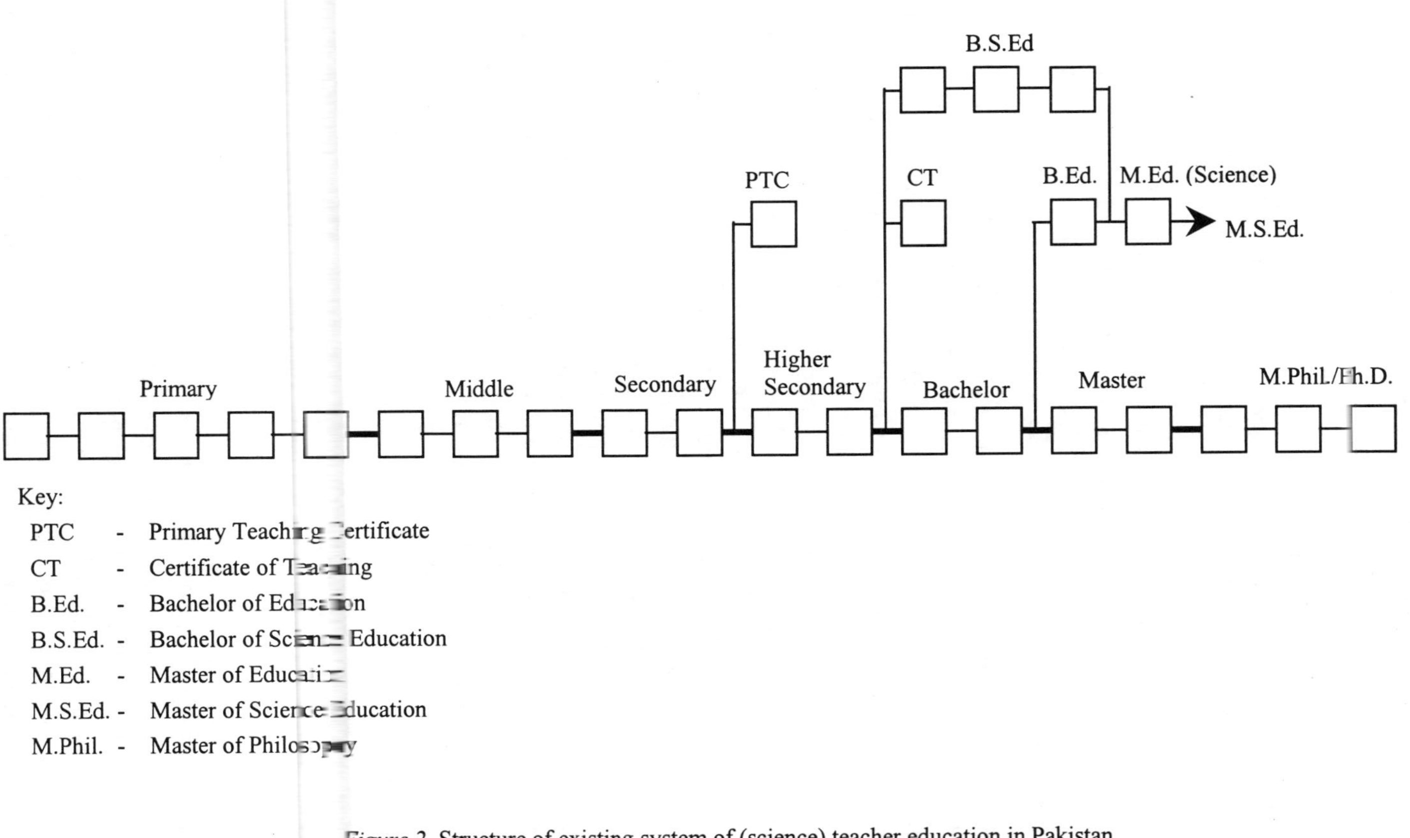

Figure 2. Structure of existing system of (science) teacher education in Pakistan

to develop science education in the country. This institute has been in operation since the 1980s, but the quality of science instruction still remains dismal. This is because the staff of the center that was to provide guidance to science teachers, though well qualified in natural sciences, lacked professional preparation in education. Some of those having professional training lacked the experience of working with school teachers and were relatively new in the field.

DISCUSSION

In the process of teaching and learning, the teacher's effort is supplemented by adequately planned curricula and a highly organized instructional methodology. The explosion of knowledge in science in the last few decades has laden the curriculum with more complex concepts, which demands an extra effort to make the teaching-learning process more effective and meaningful for both the teacher and the students. The widely recognized constructivist philosophy puts a new demand on the part of teachers to mediate the process of learning. However, all of these ideas are nonexistent in teacher preparation programs in Pakistan. Despite all-out efforts to improve teacher education programs, nothing substantial has been achieved in this regard. The teacher education program has invited criticism from the public in general, and from educators in particular. The criticism levied against teacher education programs has centered around the deficiency and inadequacy of the preservice program and inefficiency of inservice programs.

Preservice Teacher Education

The 1-year preservice program (that actually becomes a period of seven/eight months due to vacations) is considered inadequate and ineffective by many educators. The courses that prospective teachers undergo are deficient both in respect to content and duration. While basic qualifications of prospective teachers need to be raised, it is also imperative that they should be exposed to longer and more intensive professional preparation and provided with the opportunities to enlarge their vision and make in-depth study of the subjects which they will be required to teach. At present, the recommended period of PTC, CT, and B.Ed. education is 48 weeks, although in practice it is much shorter. The Commission on National Education (Government of Pakistan, 1959) recommended enhancing this period up to two years. Various other policies have shown dissatisfaction with the duration of preservice teacher education programs. But despite all these reiterations, the preparation period has not been extended. Second, there is a general criticism that a matriculate (high school graduate) receives just one year of professional preparation and is declared prepared to teach in a primary school. This immature young person, with very limited knowledge and perhaps narrow visions, is entrusted with the responsibility of guiding the destiny of the new generation. The professional preparation offered is very short as compared with other countries. In India, Iran, Korea, Nepal, and Singapore, for example, the minimum period of training for primary school teachers is two years. Malaysia prescribes three years training, China four to five years, and Indonesia six years. Thus, compared with other countries in the region, primary teacher preparation in Pakistan is insufficient.

In addition, the period of practice teaching is considered grossly inadequate. Under the existing program, the duration of practice teaching is only four weeks (Government of Pakistan, 1976). Many educators now are of the view that this period is insufficient to provide adequate practice to beginning teachers. There remains a gap between theory and practice (Iqbal & Sajida, 1997).

Thus, there is a need to introduce some fundamental change in the preservice education program for science teachers in Pakistan. It is imperative that not only the curricula of teacher education be changed, but also concomitant changes be brought about in the methodology of conducting teacher education courses. As a result of the severe criticism of teacher education, and following the recommendation by various agencies, efforts are being made to improve qualitative aspects of teacher education. Recently a Teacher Training Project (TTP) has been implemented with funding from the Asian Development Bank. Under this project, a massive In-Country Fellowship Program is being conducted for professional growth of faculty in Colleges of Education. Secondly, a change in the teacher education program has been suggested. The proposed model of teacher education is shown in Figure 3 (Government of Pakistan, 1996). However, keeping in view past experiences with teacher education reform in Pakistan, it is unlikely that these changes will readily be implemented.

Inservice Education

At the time of independence, there was no provision for the continuing education of teachers at various levels. As a result of the recommendations made by the Commission on National Education 1959 (Government of Pakistan, 1959) provisions were made for inservice education, particularly in science education. The Education Extension Center (EEC) was established,

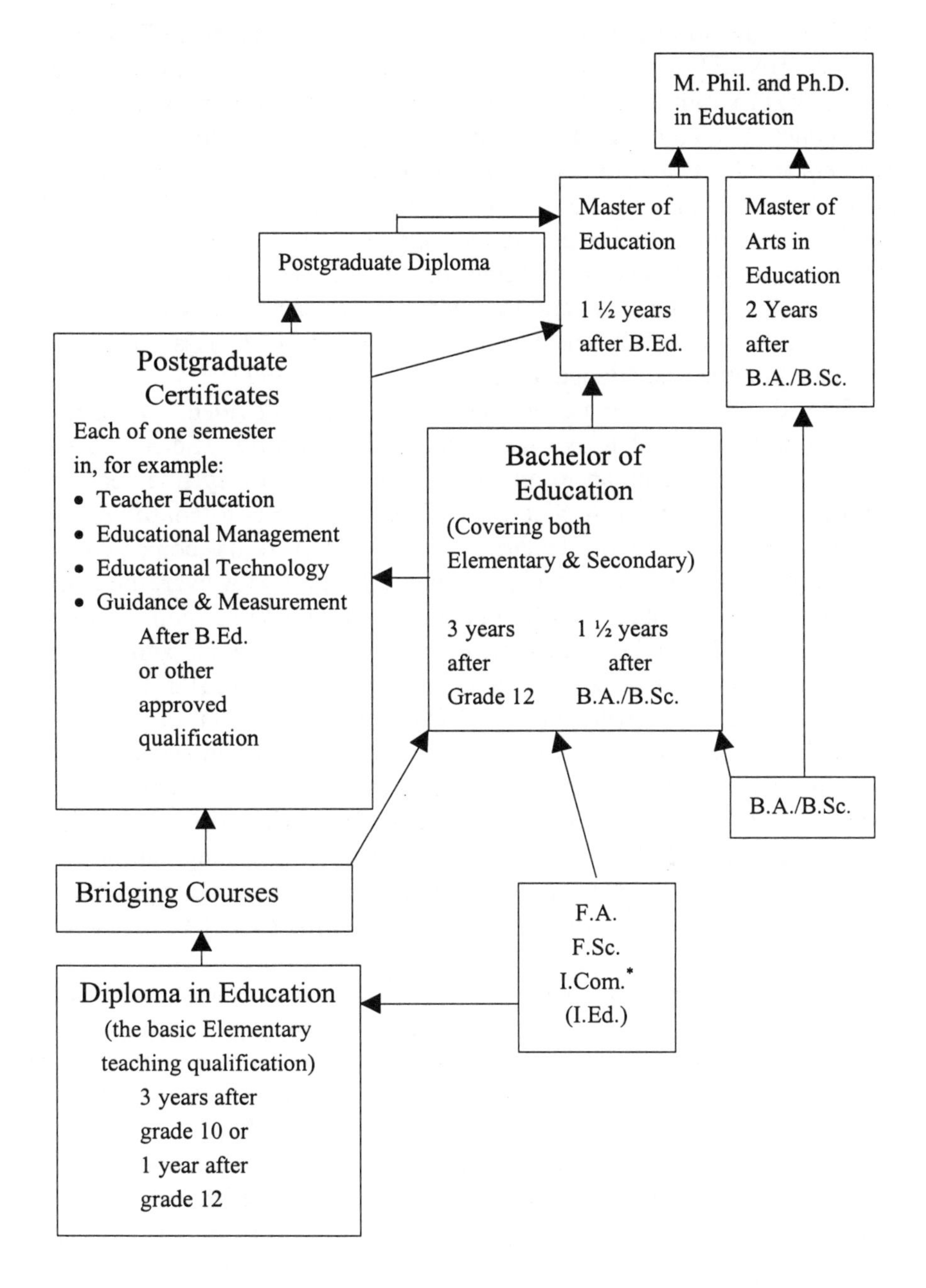

*I.Com. – Intermediate in Commerce; I.Ed. – Intermediate in Education. These represent degrees between the secondary F.A./F.Sc. and the bachelor's.

Figure 3. The proposed model of teacher education in Pakistan, 1996

mainly to cater to that need. The philosophy of the Inservice Education of Teachers (INSET) was to update the content knowledge of science teachers. A massive program of inservice education was carried out in the early 1970s. The model adopted by the EEC was based on the trickle down theory. Key Trainers at the EEC were responsible to prepare Master Trainers, who were then required to train teachers at the distant centers. A survey conducted by the EEC itself into the impact of such inservice education showed that the effect was not translated to the grassroots level. The survey showed some impact on the Master Trainers, but, when the impact to classroom teachers was assessed, the result was insignificant. Abilities of classroom teachers remained unimproved (Education Extension Center, 1977).

Thus the inservice education currently being provided in specialized institutions has proved to be inadequate and ineffective. Science teachers are required to undergo an inservice training once in every five years. The main purpose of these inservice courses in principle is to contribute to the professional growth of the teachers, improve their skills, competence and effectiveness of teaching, and change their attitudes. However, the research evidence shows that the teachers undergoing these courses come back with little improvement in their academic competence, if any at all, and no change in their attitudes (Iqbal, 1986; National Education Council, 1987).

One reason for ineffectiveness of INSET programs is that they are formal and highly structured. When planning these courses, attitudes and interests of the teachers are not taken into consideration. Under the Teacher Training Project (TTP, 1994-99), a massive program of inservice education of faculty of the Colleges of Education has been undertaken. Nothing can be said at this moment about the effectiveness and utility of this inservice program, but keeping in view the past record of such programs, expectations cannot be high.

The professional growth of teachers is not an activity that should be entrusted solely to formal institutions. Other informal means must also be adopted, such as workshops, symposia, summer courses, seminars, and meetings of professional associations. Teachers are not conscious of the need for their own professional growth and science teachers' associations are not very active. The only body concerned with the professional growth of science teachers is the Pakistan Association for Science Education (PASE). This association was established in 1990, and hence is still in its infancy. PASE has held two large-scale seminars on science education, one in 1992 and the other in 1997, with the cooperation of International Council for Associations of Science Education (ICASE). However, there is a need to hold such meetings frequently, so that teachers be given the chance of managing their own professional growth.

SUMMARY

Analysis of past developments in teacher education in Pakistan has shown that substantial progress has been made in this field. It has, however, been pointed out that education of science teachers still needs much improvement. At the present, there is an emergent need to meet the shortage of qualified science teachers and at the same time to bring qualitative improvements in the courses offered in teacher education institutions. First, we recommend that the 1-year duration of teacher preparation is grossly inadequate for all teaching courses, and should be lengthened, and the qualifications for entrance be increased. We believe that teaching must be made a graduate profession. For example, the basic qualification of primary school teachers for admission to teacher education institution should be increased. We recommend that PTC should be made a 12 + 2 year program. Similarly, CT, 12 + 3; B.Ed., 14 + 2; B.S.Ed., 12 + 4; M.A.Ed., 14 + 3; and M.Ed. one year after B.Ed. or B.S.Ed.

Secondly, we think the quality of instruction in teacher preparation programs should be improved. Most teachers in the teacher preparation institutions use the lecture method most of the time. Prospective teachers behave like passive listeners to their teachers. They do not participate in the teaching/ learning process. Some instructors even dictate their notes to the preservice teachers. When the teachers join schools, they behave the same way. Most of the courses offered in teacher education institution were designed many years ago when a behaviouristic philosophy was in vogue. The constructive approach has yet to find its way into these institutions.

Another important issue is the lack of coordination between different kinds of teacher education institutions and between these institutions and the schools. A different administration is set up for schools, colleges, and universities. Schools are administered by a separate hierarchy. Colleges of Education offering B.Ed. and M.Ed. programs come under different administrators. Universities are controlled by the Federal Government through the University Grants Commission. Departments of Education and IERs cater to the schools' need so far as the preparation of teachers is concerned. However, there is little coordination and cooperation among different agencies responsible for the professional development of teachers. This has created a wide gap between institutions offering science teacher preparation and the schools. Teacher education institutions do not receive feedback from the field and teachers do not turn to the faculty for guidance.

These are some of the issues and problems that science teacher education in Pakistan faces at present. Faculty of teacher education institutions are at the apex position so far as the teaching cadre is concerned. They are the people who are well placed. They know and should know the new developments in their field of study or profession. It is binding upon them not only to bring about changes in their own institutions, but also to introduce changes in the overall system of science education in Pakistan.

REFERENCES

Education Extension Center. (1977). *A survey of outcomes of inservice training programs.* Lahore: Education Extension Center, Government of the Punjab.

Government of Pakistan. (1947). *Proceedings of the Pakistan Educational Conference.* Karachi: Education Division, Ministry of Interior.

Government of Pakistan. (1959). *Report of the National Commission on Education.* Karachi: Ministry of Education.

Government of Pakistan. (1972). *The education policy (1972-80).* Islamabad: Ministry of Education.

Government of Pakistan. (1975). *Elementary science curricula for classes VI-VIII.* Islamabad: Ministry of Education, Curriculum Wing.

Government of Pakistan. (1976). *Curriculum outline for B.Ed. program.* Islamabad: National Committee on Secondary Teacher Education Curriculum, National Bureau of Curriculum and Textbooks, Ministry of Education.

Government of Pakistan. (1977). *Teacher education in Pakistan.* Islamabad: Curriculum Wing, Ministry of Education.

Government of Pakistan. (1979). *National education policy and implementation program.* Islamabad: Ministry of Education.

Government of Pakistan. (1981, November). *Development of education in Pakistan (1978-80).* Islamabad: Country Report for the 38th Session of International Conference of Education, Geneva, Ministry of Education.

Government of Pakistan. (1989). *National education conference: Working papers.* Islamabad: Ministry of Education.

Government of Pakistan. (1992). *National education policy 1992.* Islamabad: Ministry of Education.

Government of Pakistan. (1996). *Curriculum for M.Ed. M.A. Education, M.Sc. Education and PGCE's.* Islamabad: Teacher Training Project, Ministry of Education, Curriculum Wing.

Hayes, L. D. (1987). *The crisis of education in Pakistan.* Lahore: Vanguard.

Iqbal, M. (1986). *Evaluating inservice courses organized by the Education Extension Center.* (Unpublished Master's Dissertation). Lahore: IER, University of the Punjab.

Iqbal, M., & Sajida, R. (1998). Observation of student teaching: A reflection on preservice education program. *Proceedings of the 10th ICASE-Asian Symposium.* Lahore: PASE-ICASE.

National Education Council. (1987). *Secondary education in Pakistan: Perspective planning.* Islamabad: National Education Council, Ministry of Education.

UNESCO. (1984). *Science education in Asia and the Pacific.* A bulletin of UNESCO Regional Office in Asia and the Pacific. Bangkok: UNESCO.

SECTION III

MAKING SENSE OF SCIENCE TEACHER LEARNING

Chapter 6

Higher Order Thinking in Science Teacher Education in Israel

Yehudith Weinberger[1] and Anat Zohar[2]
[1]*Kibbutzim College of Education, Israel:* [2]*The Hebrew University, Israel*

Abstract: This chapter describes a course called *Thinking in Science* that is part of a junior high school teacher preparation program in an Israeli college. The purpose of the course is to prepare prospective teachers to integrate instruction of higher order thinking skills into science topics. A qualitative evaluation study of the course examined processes that took place, documented in a portfolio. The findings show a developmental trend in four different aspects. 1) students' ideas about instruction of higher order thinking; 2) students' opinions and attitudes regarding the course; 3) students' experiences in developing higher order thinking as learners; and 4) experiences developing higher order thinking as teachers. Learning processes during the course took place on both a cognitive and an affective level. Students' development went through a stage of cognitive imbalance, indicating meaningful learning. We also discuss the implications regarding the introduction of higher order thinking into science teacher preparation programs.

Although schools have been trying to teach higher order thinking for decades (Resnick, 1987), numerous studies indicate that they have not been very successful in achieving this goal (Dossey, Mullis, Lindquist, and Chamber, 1988; Mullis & Jenkins, 1988; 1990). An overview of science curricula and learning materials from various countries reveals predominant occupation with facts and little occupation with ways of producing knowledge (Duschl, 1990). Scientific processes that bring about new knowledge do not receive proper attention in school. A similar picture emerges from the examination of instruction in many science classrooms (Friedler & Tamir, 1984; Mendelowitz, 1996). Instruction focused on dispensing information produces students who are not proficient in higher order thinking skills in general or in scientific inquiry skills in particular.

S.K. Abell (ed.), Science Teacher Education, 95–119.

However, recent educational endeavors show that when interventions are explicitly directed towards fostering students' thinking, they can bring about significant improvement (Brown & Campione, 1994; Bruer, 1993; Feurstein, Rand, Hoffman, and Miller, 1980; Lipman, 1985; Shayer & Adey, 1992a; 1992b; Zohar, Weinberger, and Tamir, 1994). As more and more "thinking" projects are being implemented in schools worldwide, it becomes clear that a serious impediment to broad and successful implementation is the lack of adequate methods for preservice and inservice staff development in this particular area.

We have every reason to assume that the instructional model that was experienced by most preservice teachers when they were school children did not emphasize learning of higher order thinking. Based on the common saying that teachers teach in a way that reflects the ways they were taught as school children, this situation does not predict that preservice teachers will apply instruction of higher order thinking without specific preparation. In addition, several studies show that science content courses within preservice programs are often based upon lectures and transmission of knowledge (De Rose, Lockard, and Paldy, 1979; Donnellan, 1982; Yakoby & Sharan, 1985). Therefore, preservice teachers may get the undesirable message that a transmission-of-knowledge approach is appropriate for instruction. The need to correct this undesired message requires a course that will highlight ways to integrate higher order thinking skills into instruction (Casey & Howson, 1993). This chapter describes a course called *Thinking in Science* that is part of a preservice program for prospective junior high school science teachers in Israel. The goal of the course is to prepare the preservice teachers for implementation of a project designed to enhance higher order thinking in science classrooms.

Integrating the development of higher order thinking into science teacher education is based on two assumptions: a) preservice teachers need to improve their own thinking abilities (Brownell, Jadallah, and Brownell, 1993); and b) exercising thinking skills will contribute to the prospective teachers' ability to advance their students' thinking skills (Bransky, Hadass, and Lubezky, 1992; Krombey, 1991; Udow, 1991).

Several researchers have proposed theoretical frameworks for the study of teachers' cognition (Clark & Peterson, 1986; Schön, 1983; 1987). Shulman's (1986) categorization of teacher knowledge, including subject matter knowledge and pedagogical content knowledge, illuminated the relationship between various categories of teacher knowledge and teaching. Shulman's initial categories have been further developed and refined in later studies (e.g., Adams & Krockover, 1997; Grossman, Wilson, and Shulman 1989; Zohar, no date). The present study accepts the general framework of Shulman's classification of teacher knowledge, adapting it to the special circumstances of teaching higher order thinking skills. Subject matter knowledge in this context is knowledge of thinking processes, and the pertinent pedagogical content knowledge includes instructional means for teaching higher order thinking. We will address these two types of knowledge, documenting their development in the *Thinking in Science* course.

In the next sections, the cultural context and national policies of science education programs in Israel will be reviewed, followed by some description of Israeli projects aimed at instruction in higher order thinking skills. Then, the course *Thinking in Science* will be described. Finally, a qualitative evaluation study of the course will be reported, highlighting learning processes that took place during the course.

CULTURAL CONTEXT AND NATIONAL POLICIES

The changes in policy towards science education in Israel have essentially paralleled those in many other countries: disciplinary reform in the 60s, interdisciplinary reform in the 80s, and socio-technological reform in the 90s.

A few years after the Holocaust, Jews from all over the world immigrated to Israel, which finally became an independent state in 1948. Under those circumstances, love of the Jewish homeland was considered a central educational value. Learning about nature in Israel was seen as an educational means towards achieving the important goal of instilling love of the homeland and strengthening the bond between the people and their land (Dressler & Levinger-Dressler, 1996). During that period, the dominant approach to science education was naturalistic-romantic.

In 1953, the National Education Board was established and a single science curriculum titled "Nature and Agriculture" was written. That curriculum included the following aims:

> Knowing the laws of natural phenomena, scientific observation methods and scientific thought… knowledge of the nature of the homeland… fostering an intimate relationship with the land and its wildlife… fulfilling the dream of agricultural labor and rural life as a valuable lifestyle...fulfilling the goal of pioneering building of the homeland. (Ministry of Education and Culture, 1954, p. 1)

In the political context of that period, agricultural development was thought of as a prominent way for fulfilling the value of loving the land. The most important characteristic of that generation was the emphasis on teaching bodies of information, focusing on facts, and emphasizing their applied and practical aspects.

In the 1960s, following the launching of the first Russian Sputnik into space, echoes of curricular reforms that took place around the world reached Israel. Translated materials of the American BSCS and the English Nuffield programs became prevalent curricula, emphasizing inquiry as a way of teaching and learning. The curricula of that period expressed the desire to produce excellent scientists who would contribute to the advancement and development of Israel. However, it is important to note that some of the goals of those programs were only partially achieved. For example, although the Israeli matriculation exam in biology demands comprehensive scientific inquiry skills (Tamir, 1985), it seems that a large proportion of biology teachers in high school teach scientific research processes technically rather than meaningfully. This is done by focusing on teaching algorithms for succeeding in the exam instead of focusing on thinking processes (Friedler & Tamir, 1984; Mendelowitz, 1996; Zohar, Schwartzer, and Tamir, 1998).

The 1980s brought about a change in the aims of science education: it was no longer seen as the initial training of future scientists but in terms of "science for all." People in Israel, like in many other modern countries, are exposed daily to circumstances that require scientific knowledge and technological-scientific literacy. This new reality, together with students' tendency to avoid studying science, gave rise to the requirement that science teaching be anchored in social contexts, emphasizing the relevance of science to everyday life. The curricula of the 1980s emphasized STS (Science, Technology and Society). STS programs diminished the role of inquiry learning as the predominant method; inquiry became only one of several suggested instructional methods from which teachers could choose.

At the beginning of the 1990s, a national committee, The High Committee for Science and Technology Education, was formed to examine the state of science education in Israel. The result of the committee's work was the Harary (1992) report, "Tomorrow 98," which has since guided the basic principles of science education in Israel. The essence of its suggestions is that:

> Science and technology education is the core of scientific infrastructure...The government of Israel will announce a national curriculum to strengthen, deepen and improve the learning preparing the next generation of citizens for life in the techno-scientific age. Implementation of this policy includes an interdisciplinary approach to the subject...science and technology in our time are inter-linked and overlap in a variety of surprising ways... the learning of science and technology must be combined. (pp. 3-4)

Computers are seen as "a most important instructional tool for any subject and for all age groups" (pp. 76-77). A fair amount of space in the report is allocated to the issue of developing students' thinking and problem-solving skills:

> In many places in the world today there are programs designed to improve the individual's creative thinking, inventive thinking, logical thinking etc...This issue is worthy of exploration. The intention is to investigate the feasibility of including such programs in our schools. (p. 47)

Likewise, the report emphasizes the central role of the teacher in science and technology education. "The best curricula and the most equipped laboratories will not bear fruit without good teachers. In the end, education stands and falls according to the quality, skills and dedication of teachers" (p. 10).

The STS approach (which characterized the programs of the 1980s) is validated in the 1990s as one of the recommendations of the Harary report. As a result of the Harary report, new curricula for science education were published. In junior high school, for example, a new curriculum was published in "Science and Technology Studies in Junior High School" (Ministry of Education, Culture and Sport, 1996). Its main aims are to educate the citizens of the next century towards technological-scientific literacy, and to prepare the background for later studies in high school. The main characteristic of this curriculum is its interdisciplinary approach, which integrates various scientific disciplines with technology in social contexts and emphasizes instruction of learning and thinking skills. "Students should be involved in the designing, carrying out, analyzing, drawing conclusions, discussing and assessing findings or solutions...in a wide range of topics in science and technology" (p. 15).

THINKING DEVELOPMENT PROJECTS IN ISRAEL

There are currently a number of Israeli educational enterprises in the field of teaching higher order thinking. We will first review two central activities in the field and then focus on the project which is at the heart of this chapter.

One of the oldest and most wide-reaching projects in Israel is the program "Instrumental Enrichment," whose aim is to improve the learning ability of the individual through developing his thinking skills (Feurstein, 1991; Feurstein et al., 1980). The program has been translated from Hebrew into many other languages and is currently implemented in many countries.

Another major enterprise in this field is the "Branco Weiss Institute for the Development of Thinking," which was established in 1990 with the aim of developing the thinking of children in the Israeli educational system. The Institute develops and produces teaching and learning modules for teachers and students; publishes a quarterly distributed to all schools in the country, as well as to private subscribers; and translates into Hebrew books on critical and creative thinking, intelligence, and related topics. The Institute also runs education programs and courses for teachers, principals, tutors, supervisors and other educators, as well as thinking clubs for children (Vinner, personal communication, 1997).

This chapter focuses on another Israeli project called "Thinking in Science Classrooms" (TSC) that was established as part of the Harary reform. The TSC project emphasizes the integration of higher order thinking skills into the science curriculum. The goal of the project is to design learning activities that aim to foster higher order thinking skills according to the infusion approach to teaching thinking (Ennis, 1989). The contents of the learning activities match topics from the regular science syllabus, so that teachers may incorporate the learning activities in the course of instruction whenever they teach a topic covered by one of these activities. The project's goal is that a set of opportunities calling for "thinking events" take place in multiple science topics. The activities are designed to foster the growth of both scientific concepts and scientific reasoning skills. The emphasis on skills does not mean that skills are taught as context-free entities. Instruction always begins with concrete problems (regarding a specific scientific phenomenon) that students are asked to solve. After students have used the same reasoning skill in various concrete contexts, they are encouraged (usually through class discussion) to engage in metacognitive activities that include generalization, identification of skills, and formulation of rules regarding those skills. In order to avoid fixed patterns of learning activities (which might eventually train students to deal with problems merely in an algorithmic way), varied types of learning activities were designed (Zohar & Weinberger, 199[illegible]): a) inquiry and critical thinking skills learning activities; b) investigation of microworlds; c) learning activities designed to foster argumentation skills about bioethical dilemmas; and, d) open-ended inquiry learning activities.

Within the TSC project, inservice and preservice staff development courses are conducted with the aim of educating teachers to implement the TSC methods and approach in their classrooms. Below we describe and analyze a preservice course, *Thinking in Science*, that is taught to prospective junior high school science teachers in a large college in the center of Israel.

THINKING IN SCIENCE: COURSE DESCRIPTION

Researchers report that educating preservice teachers in critical thinking and inquiry teaching may result in both improved attitudes and improved thinking skills (Sesow, 1991). It therefore seems worthwhile to introduce issues that involve teaching and learning of higher order thinking skills into preservice science teacher education. Previous research also shows that introducing changes in teacher behavior in general, and changes specifically geared towards using more inquiry-oriented teaching approaches, is definitely more complex than originally thought (Adams & Krockover, 1997; Casey & Howson, 1993). Such desired changes require that teachers not only learn new facts, but also rethink what they already know. In order to change, teachers need to adopt new knowledge and desired practices related to teaching (Adams & Krockover, 1997; Hewson, Kerby, and Cook, 1995). Therefore this complex change process requires well-designed and focused education programs. The design of the *Thinking in Science* course described below was aimed at that goal.

Purpose

The purpose of the course is to address the following issues:

1. To discuss the importance of fostering students' higher order thinking skills in science lessons.
2. To review several projects and /or curricula designed to foster higher order thinking.
3. To improve preservice teachers' higher order thinking skills and their awareness of metacognitive processes.
4. To consolidate preservice teachers' perceptions regarding instruction of higher order thinking in science.
5. To introduce the TSC learning materials.
6. To advise preservice teachers in planning ways of integrating those learning materials into their practical work in science classrooms.

Course Structure

The course consists of three basic components (see Table 1) that are included in each unit:

1. Mini lectures and class discussions. This component of the course includes several general theoretical issues regarding instruction of higher order thinking. The main issues are: a) definition and clarification of concepts regarding instruction of higher order thinking; b) the rationale for integrating instruction of higher order thinking into science lessons;

c) review and analysis of various projects and/or curricula designed to teach higher order thinking; d) cognitive aspects of thinking; e) developmental stages of thinking strategies; and f) assessment of higher order thinking.

2. Active practice. This includes experience with a wide variety of learning materials (taken from the TSC project). Preservice teachers solve problems presented in the learning materials, analyze their logical structures, analyze typical difficulties that children encounter while they interact with those learning materials, and think about appropriate means of instruction.
3. Reflective practice. This component of the course addresses metacognitive processes. Students reflect upon the thinking skills they applied while engaged in solving the TSC problems and upon their own learning processes.

The way in which the three parts of the course are combined is described in Table 1. During the first part of the course (lessons 1-5), the mini lectures and discussions component are predominant. During the middle part (lessons 6-9), active practice is highlighted, and during the final part (lessons 10-14), the emphasis is upon reflective practice.

Sample Lesson Plan

In order to demonstrate how the three components are combined in a specific topic, we will describe a sample unit that was taught at the middle of the semester (lessons 7 and 8). This unit revolved around one of the TSC learning activities which investigates factors that may influence the rate of seed germination. The learning activity for the students included a computer simulation, a set of worksheets, and a video of a pupil working with the learning activity (Zohar, 1996). The unit included the following stages:

1. Active practice: Students solved the problems presented in the learning activity (in the same way that school pupils usually do).[1] The goal of this stage was to let preservice teachers get acquainted with the learning activity.
2. Reflective practice: Students reflected upon their own thinking processes (that took place during the problem solving stage), focusing on analyzing the thinking skills they had used.
3. Active practice: Students were then prompted to think "as teachers". First, they were asked to predict pupils' difficulties while solving the

[1] In order to avoid confusion between college students and school children, we refer to college students as "students" and to school children as "pupils."

Table 1. Three Basic Components in *Thinking in Science* Course

Lessons 1 - 5	Lessons 6 - 9	Lessons 10 - 14
Reflective practice unit Reflection on instructional processes that took place during instruction.	Reflective practice unit Writing reflection number 1. Discussing: meaningful learning/teaching.	Reflective practice unit Writing reflection number 2. Assessment of the portfolio. Reflecting on conceptual change. Assessment of the learning and thinking processes.
Active practice unit Analysis of a case "Harry S." "Melinark" activity. "Thiamin in rats" activity.	Active practice unit "The particular structure of the matter" activity. "Microworld" activity. Logical analysis of the microworld. Pupils' confrontation with the microworld. "Water in Living Organisms" activities. Critical reading of articles and commercials.	Active practice unit "There is no hole in Ozone" activity. "Zemmelweiss" activity. Designing HOT learning activities. Reading an article about instruction of HOT.
Mini lectures & discussion unit Educational purposes. Psychological aspects of HOT. Definitions of HOT concepts: "Generative Knowledge" "Knowledge" "Information" Rationale of integrating HOT into instruction.	Mini lectures & discussion unit Various approaches for instruction of HOT. Stages in acquisition of new thinking strategies. Assessment of HOT.	Mini lectures & discussion unit Principles of instruction of HOT.

HOT = Higher Order Thinking

problem. Then segments of the video were shown, and students were asked to diagnose the pupils' difficulties and propose how they would continue their lesson if this pupil was in their classroom. Students first thought of those issues independently and later shared their thoughts with the whole class.

4. Mini lectures and class discussion: The instructor led a class discussion about diagnosis and treatment of pupils' thinking difficulties. Then she supplemented the ideas that came up in class with a theoretical mini lecture about the development of thinking strategies (Kuhn, Garcia-Mila, Zohar, and Anderson, 1995; Siegler & Jenkins, 1989). Finally, the instructor presented all the learning materials (including worksheets) that were prepared by the TSC team for that learning activity, and discussed their rationale.

EVALUATION RESEARCH

The course *Thinking in Science* has been accompanied by an evaluation research. The full evaluation study consists of three parts:

1. The first part is a quantitative research addressing preservice teachers' prior knowledge and dispositions towards thinking.
2. The second part documents and analyzes the learning processes that took place throughout the course.
3. The third part examines the degree to which the content of the course is reflected in the practical teaching of four preservice teachers.

Part 2 of the study and some of its findings are described in this chapter.

Participants

Participants were students in a preservice program for junior high school science and technology teachers. The 4-year program includes courses in science, education, and science education, as well as extensive guided field work in schools: a group of two to four students collaborates in teaching one science class (for two consecutive years). The program grants a bachelor's degree in education (B.Ed.). The students who participated in the *Thinking in Science* course were in their second or third year of the program. There were 22 females and two males. Their ages ranged between 22-29 years.

Student Portfolios

Preservice teachers' work during the course was documented in a portfolio that reflected their achievements or progress. The portfolio for the *Thinking in Science* course followed the principles of Arter and Spandel (1992). In order to avoid a random collection of materials, the following specific materials were selected:

1. A collection of documents, sampling students' written work. The collected documents included questionnaires, assignments, and products of creative workshops. Data from assignments in which the same issues were addressed at different times were useful for portraying developmental trends throughout the course.
2. At least one additional document chosen by each student was included, in order to allow individual expression of student development.
3. Individual reflections were written on two different occasions at the middle and at the end of the semester. In these reflections students were asked to:

   Explain in detail your views and opinions about: 1) instruction in higher order thinking; 2) learning higher order thinking; 3) the course *Thinking in Science*; 4) processes you went through during the course; and 5) anything else you may care to write about.

Management of the portfolio. The portfolio was owned by the student. Therefore he/she was responsible to collect all relevant materials and to turn them into the teacher for assessment. Particular tasks were handed to the teacher throughout the semester. The whole portfolio, including students' self assessment, was turned in at the end of the semester.

Assessment of the portfolio. The portfolio was used as the only means of grading students' work. The quality of students' work was assessed according to the following criteria:

1. Perceptions regarding instruction of higher order thinking.
2. Mastery of higher order thinking, including procedural and metacognitive knowledge.
3. Mastery of instructional means appropriate for teaching higher order thinking.
4. Ability to design new learning materials aimed at instruction of higher order thinking integrated into science content.

The portfolio was jointly assessed by the student and the teacher. First each student assessed her portfolio based on given guidelines (see Table 2). Then the teacher assessed the portfolio using the same criteria. Finally the teacher and the student met to discuss and compare the two assessments as well other topics relevant to the course raised by either student or teacher.

Table 2. Guidelines for Portfolio Assessment

1. Organize the documents in the portfolio in chronological order. The content of the portfolio reflects your progress and achievements during the semester.
2. The "evidence" in the file should allow you to characterize the processes that you have undergone in at least four areas:
    - Perceptions regarding instruction of higher order thinking.
    - Mastery of higher order thinking skills.
    - Mastery of instructional means appropriate for teaching higher order thinking.
    - Designing new learning materials suitable for fostering students' higher order thinking.

    a. Describe and assess your progress in each of the areas listed above as demonstrated by the contents of the portfolio (consider each section separately).

    b. In summary, make a short assessment of your work and achievements in the course as demonstrated by your portfolio.
3. Write a reflection summarizing the topic "Developing Thinking in Science Teaching." Relate to any aspect you consider relevant (do not worry about repeating things you have already mentioned elsewhere).
4. Hand in your assessment together with the portfolio itself by the due date. Within two weeks of handing in your work, the instructor will meet with you to discuss the portfolio and jointly assess it.

Documentation and Analysis of Learning Processes

Evidence of development in students' thinking during the course are based on three sources:

1. **A questionnaire about fostering higher order thinking in science education.** The questionnaire was given in the first session of the course in order to reveal students' ideas and opinions prior to the course. The questionnaire presents briefly two approaches as to the aims of science instruction in schools and asks students to relate to them. The first approach emphasizes the delivery of a wide range of information on science related topics, the second emphasizes the development of scientific thought. Students were asked to indicate arguments for and against each approach and indicate which approach they supported and to justify their choice (to take a stand).
2. **First and final reflection.** The first reflection was written in the sixth lesson of the course, the final at the end of the semester.
3. **Overall portfolio assessment.** At the end of the semester, students and instructor assessed student portfolios.

Characterization of processes that took place during the course was carried out by analyzing the questionnaire, reflections, and other portfolio documents according to the contrast/comparative method (Miles & Huberman, 1994). First, we read the documents found in the portfolios thoroughly, and jotted down ideas about repeating themes. Then, we separated those documents into several different aspects according to the following themes: a) ideas about instruction; b) opinions and attitudes regarding the course; c) experiences in the development of higher order thinking learners; and d) experiences in the development of higher order thinking as teachers.

The unit of analysis was a single statement. First, in each activity, statements were categorized and counted. In the second stage, the number of statements in each category and the categories themselves were compared in different periods during the semester, in chronological order.

When reading the findings below, it is best to bear in mind the plan of the course (depicted in Table 1) so that the content of students' statements at the middle and at the end of the course may be compared to the overall course plan.

FINDINGS: THE DEVELOPMENT OF STUDENTS' ATTITUDES AND KNOWLEDGE REGARDING INSTRUCTION OF HIGHER ORDER THINKING IN SCIENCE EDUCATION

This section describes students' learning processes expressed by their written work[2] at three separate times during the course: at the beginning, the middle and the end (see Table 3). Four different aspects were considered:

1. Ideas about instruction of higher order thinking.
2. Opinions and attitudes regarding the course Thinking in Science.
3. Experiences in developing higher order thinking (as learners).
4. Experiences in developing pupils' higher order thinking (as teachers).

Analysis of students' ideas about instruction of higher order thinking was based on three sources from the portfolios: the questionnaire given at the beginning of the course, reflection number one, and reflection number two. Analysis of students' ideas regarding the other three aspects was based on the two reflections only.

Ideas About Instruction of Higher Order Thinking

At the beginning of the semester, 12 students' references to the issue of higher order thinking instruction were tautological; for example: "It's important to teach science together with development of thinking, because thinking is a very important issue in instruction." Others' ideas about instruction of higher order thinking skills were limited to two topics: a) the importance of developing students' thinking skills: "Developing one's thinking is a tool that will remain with the student forever and will help her to cope with challenges and to acquire knowledge in various fields;" and b) reference to the explosion of information that is typical of today's culture: "Knowledge accumulates very fast so there is no way we can teach it all." It should be noted that all students' arguments at this stage were general theoretical statements about education and not practical ideas that could be used in teaching.

At the beginning of the course, the attitudes of 14 students towards instruction of higher order thinking in the context of science education was found to be positive. However, this finding should be treated with caution because of social desirability. The students had answered this questionnaire in the first lesson of a course entitled *Thinking in Science* and so may have accommodated their responses to please the lecturer.

[2] It should be noted that the data excerpts were translated from Hebrew by the authors.

By the middle of the semester, students' knowledge and ideas were more sophisticated and multi-dimensional in comparison with the beginning of the semester. In addition to the ideas that appeared before, new ideas started to emerge. Some students displayed usage of professional terminology, such as:

1. References to a child-centered educational approach:
   Instruction of higher order thinking is focused on teaching thinking skills, creativity and open mindedness. It encourages acquisition of learning skills such as individual work, group work and team work, i.e., it emphasizes child development in a child-centered approach.
2. References to a constructivist approach:
   Instruction of higher order thinking is based upon active construction of knowledge by the students who think and solve problems.
3. References to the idea of "generative knowledge" (Perkins, 1992):
   Instruction of higher order thinking helps to internalize knowledge, to acquire meaningful knowledge which can then be used in the future, to organize knowledge and to connect it to prior information.

Another new point emerging at this stage of the course was that students started to bring up issues related to instructional aspects of higher order thinking:

Instruction according to the TSC approach is appropriate for the new science curriculum and improves teaching.

In this context, students noted the contribution of the course to the professional development of the prospective teacher and to changes in teachers' conceptions and work habits:

Instruction according to the TSC approach necessitates a change in the ideas and in the thinking methods of the students and require teacher education.

It is especially important to point out that, at this stage of the course, nine students expressed negative attitudes, noting difficulties and drawbacks in teaching higher order thinking. Students thought that teaching thinking is expensive, requires a lot of time, will be at the expense of covering the curriculum, and that instruction of higher order thinking may cause difficulties because it requires radical changes in teachers' thinking and working habits.

At the end of the semester, students' knowledge and ideas about instruction of higher order thinking were more solid and multi-dimensional compared to the middle of the semester. Students' ideas by the end of the course were diverse and included many different aspects:

1. General education rationale:
   It is important to incorporate instruction of higher order thinking to science lessons, because it will prepare pupils for the future and will develop their skills and abilities.

Table 3. Summary of the Results

At the beginning of the semester	At the middle of the semester	At the end of the semester
Ideas about instruction of HOT	Ideas about instruction of HOT	Ideas about instruction of HOT
One reference and only a few aspects. Difficulties in expression and formulation of ideas. Positive attitudes towards instruction of HOT. General theoretical statements. Reference to the learning aspect.	2-3 references to multi-dimensional aspects. Improvement in formulation of ideas with usage of professional terminology. Appealing the positive attitudes toward instruction of HOT. General theoretical statements. Reference to the teaching aspect.	3-4 references to four dimensions. Usage of correct professional terms in correct contexts. Reference to the professional demands required from teachers. Positive attitudes towards HOT instruction. General theoretical and practical statements. Reference to principles of instruction of HOT.
	Opinions and attitudes regarding the course	Opinions & attitudes regarding the course
	Some mentioned contributions of the course and some absence of contribution.	Detailing of the contributions of the course.
	Experiences in developing HOT as learners	Experiences in developing HOT as learners
	Lack of experience with HOT and expression of the need.	Positive experiences in developing HOT.
	Experiences in developing HOT as teachers	Experiences in developing HOT as teachers
	Dilemmas and fears in implementation of the approach. Few implementations of the approach in instruction.	Reports of using the TSC approach in teaching, applying TSC learning materials, changing the way of teaching, tracing pupils' improvement in HOT.
	Awareness to processes	Awareness to processes
	Gaining knowledge in thinking and teaching. Awareness to thinking processes. Awareness to learning processes. Vague concepts of the course.	Growing awareness to processes they underwent during the course: a) understanding the essence of the course; b) awareness of the learning skills; c) changing ideas about teaching science; d) metacognitive processes; and e) empowerment.

HOT = Higher Order Thinking

2. Rationale related to instructional methods:
   Using the TSC approach improves learning processes and internalization of information... induces active learning... increases motivation and challenges pupils.
3. Science education rationale:
   The TSC approach is suitable to the spirit of the new science curriculum...to teaching scientific thinking skills... to teaching procedural aspects of science.
4. Principles of instruction of higher order thinking:
   The TSC approach creates a learning atmosphere that is characterized by openness to different forms of thinking. The TSC learning materials create opportunities to practice thinking skills and engage in metacognitive activities.

This last aspect is a striking addition to the ideas that appeared in the middle of the semester. Specific principles of instruction that are of practical nature bear testimony to an additional level in students' knowledge:

Instruction according to the TSC approach requires perseverance in letting pupils exercise their thinking skills, in order to improve their thinking.

In teaching according to the TSC approach it is important to use metacognitive processes in order to generate pupils awareness of thinking skills on a general level.

In addition, in most of their responses, students had used correctly professional terms that were studied in the course and had applied them in correct contexts:

In teaching according to the TSC approach, the teacher navigates her pupils and leads them by organizing a learning environment that will allow pupils to experiment thinking and to construct knowledge.

At this stage all the negative feelings that were articulated by the middle of the semester expressing concerns regarding the TSC approach had disappeared. Instead, the opposite attitude was stated:

Teaching according to the TSC approach is not necessarily instead of teaching subject matter.

Some students commented on the professional demands required of teachers:

Instruction according to the TSC approach demands from the teacher much thought and investment during lesson preparation and during teaching. In addition she needs to be proficient in the theoretical, instructional and content-knowledge aspects of the subject.

Such statements reflect student knowledge about the professional tools which are needed for effective teaching according to the TSC approach.

These findings clearly demonstrate the development of students' ideas during the course regarding instruction of higher order thinking.

Students' Opinions and Attitudes Regarding the Course *Thinking in Science*

In the middle of the semester, 14 students wrote that the course had contributed to their own individual general knowledge and/or to their instructional abilities:

The course conceptualized and arranged knowledge that we had before.

Through the course I became familiar with different teaching and learning methods.

The course makes me look at things I learn from a different angle.

I have learned to recognize different ways of thinking.

Through the course I have gained useful tools for developing thinking – both my own thinking and my pupils.

The course has clarified and sorted out things I already knew.

The course is a preparation for practical work: it has given me ideas about how to organize my teaching, it has exposed me to practical ways of using the TSC approach and has pointed out areas to watch for in the course of instruction.

However, six students expressed difficulties and concerns about whether or not there were any meaningful benefits to the course:

I don't know if I gained much from the course.

I don't see much connection between the theoretical side we have learned in the course and field implementation.

Evaluating the course, at the middle of the semester, students emphasized that the course was interesting and important:

The lessons were enjoyable and interesting.

Everyone should take this course.

The reflections written at the end of the course, as well as the portfolio assessment, show that students thought the course contributed to them in:

1. Revealing new ideas:

   The whole issue of developing students' thinking was only revealed to me for the first time in this course. It had opened up for me a whole different way of working with pupils.

   During the course we were exposed to many new ideas and opinions.

2. Raising awareness of their own learning and thinking processes:

   I feel that this [the course] helped to the crystallization of new knowledge within me.

   The course helped me in that it exposed me to the subject and made me more sensitive to it.

During the course, with the help of a variety of activities, some order has been established `amongst the mess` I had in my mind before and now I am able to distinguish between the different thinking skills.

Now that I look back, I can see that it's not that my own thinking was so deficient, it is just that I was not aware of everything that I am aware of today.

Another difference is that during the course I was exposed to far more opportunities to think.

3. Improving their competence in instructional means for teaching higher order thinking:

   I believe that a great advantage of the course is that it made me understand that thinking should not be taken for granted, that you must interfere (with regular teaching) and explain about thinking skills, discuss each specific thinking skill and understand that both adults and children have difficulties applying it.

Some of the references were similar to those from the middle of the semester, although they were somewhat richer or more profound. However, the issue of students' awareness of their own learning and thinking processes was added at this stage. In general, the course was described as a process that demanded "thinking and practical ability," gave a lot of practice in thinking, and provided feedback. The students emphasized that the course was important and contributed to their education as teachers. Five students pointed out that the course should have been continued for another semester:

The course was too short. It is a pity that we could no carry on for another semester.

In summary, our findings by the end of the semester indicate that students thought that the course was important for their professional development and that they were aware of some specific aspects of its contribution.

Experiences in Developing Higher Order Thinking (As Learners)

Student reflections in the middle of the course indicated that the classes they had taken in college did not involve higher order thinking:

Developing thinking requires practice and personal experience. In college most classes are taught frontally and developing thinking is not encouraged.

Students were critical towards that situation, expressing the opinion that:

The college needs to emphasize the development of higher order thinking in its science courses as well.

In addition, students wrote that practice would improve their ability to teach according to the method in the future:

> *Learning [according to the approach of] developing thinking skills will enable us to teach according to that approach in our classrooms.*

In the reflections written at the end of the semester as well as in their portfolio assessments, students made fewer references, compared with the middle of the semester, to experiences in developing higher order thinking as learners (9 versus 19 references respectively). In the middle of the semester, students referred mainly to the absence of instruction of higher order thinking in the college and the associated disadvantages. Now, at the end of the semester, students referred mostly to their positive experiences in the TSC course and its contribution to their learning:

> *Exercising and practicing higher order thinking skills during the course helped me to internalize that approach and to use it in my teaching.*

According to the students' reports, the course gave them basic practice and experience that would enable them further development in this area:

> *I now have a good foundation in thinking development but I still need to learn more and improve my own ability as well as my teaching abilities regarding this method.*

Students mentioned that their success resulted from the practical experiences they were engaged in during the course. In summary, our findings by the end of the semester indicate that students experienced a development in their own thinking.

Experiences in Developing Pupils' Higher Order Thinking (as Teachers)

Students' reflections in the middle of the semester indicated that only five students experienced instruction of higher order thinking in their practical fieldwork. Those experiences were described in several different ways:

1. A structured effort to make pupils think:
   During the lesson I make my pupils think in a structured manner, as I learned in the course.
2. A change in teaching methods:
   Teaching according to the TSC approach influences the way I teach (non-frontal).
3. Creating an atmosphere of openness that may foster thinking:
   While teaching according to TSC approach, the atmosphere in the class is different [than in other lessons].
4. Awareness of instructional means for teaching thinking:

When I am preparing my lesson plan, designing teaching materials or observing others teaching or preparing projects, I think of how I can develop pupils' thinking.

In contrast to this positive feedback, however, one third of the students (seven students) expressed dilemmas or concerns regarding instruction of higher order thinking. One of their dilemmas stemmed from a difficult choice they believed they must make between two central educational values: transmission of knowledge versus fostering higher order thinking:

I have a dilemma with regard to teaching by this method: what is more important, covering the material or developing thinking? It seems to me that we have to combine the two and to adapt them in accordance with the level of the class.

Dilemmas like these did not appear at the beginning of the course, even though the questionnaire "invited" references of this kind. Other concerns raised in the middle of the course revolved around the difficulty of changing pupils' and teachers' existing thinking patterns and the need for support in the application of the theoretical principles which were learned in the course:

I have a dilemma about teaching by this method; I am not yet sure that I will be able to apply what I have learned within the framework of my teaching at school.

At the end of the course, students' reports showed an increase in the number of students who reported using the TSC approach with their pupils compared with the middle of the semester (15 versus 5 students respectively). Some of the instructional means that had been used included an application of the TSC learning materials. However, students also reported that they introduced thinking to their pupils in non-structured ways by asking higher order thinking questions when leading whole class discussions and by using unexpected opportunities that occurred during instruction to practice thinking:

During my lessons, I make my pupils think in a structured manner (as I learned in the course). I use suitable learning materials, include aspects of thinking development in whole class discussions and try to expose pupils as much as possible to thinking processes. I try to let my lessons flow and I improvise in accordance to pupils' ideas and reactions.

This quotation indicates that the open-mindedness and flexibility required of teachers engaged in instruction of higher order thinking, had been transferred from a theoretical to a practical level. Likewise, ten students wrote that the developing thinking method changed the way they taught:

Teaching by the TSC approach has changed the way I teach. I make greater use of worksheets, the pupils are more active and I create an atmosphere conductive to practicing thinking skills.

Students reported, too, that they were aware of the issue of thinking development when they prepared their lessons:

> *When I prepare my lesson plan, create teaching aids, or when I observe other teachers, I think how I can develop pupils [thinking].*

Students' references at this time began to include unique descriptions of their classrooms as they taught according to this method. For example, some traced their pupils' improvement in thinking ability:

> *During instruction it is interesting to watch the way pupils' thinking develops and improves, to listen to their answers and watch as they succeed to solve problems.*

These data provide evidence that students applied the TSC principles in their practical fieldwork. Again, the concerns and dilemmas that we witnessed at the middle of the semester were no longer found at the end of the course. These findings clearly demonstrate development of students' experiences in higher order thinking as teachers.

CONCLUSIONS

Based on our findings, it is possible to describe students' development during the course. At the beginning of the course, students' attitudes towards instruction of higher order thinking were positive. Their ideas at this stage were based on general knowledge or on social desirability, and not on specific knowledge. Many of them had difficulties expressing their ideas. As the course continued, two developmental trends became apparent.

First, during the course, students gained new knowledge about instruction of higher order thinking. This knowledge was not only theoretical, but also included practical aspects of instruction. With time, students' ideas became more complex and more sophisticated. Students reported an increase in using the TSC method in their practical work with a growing feeling of confidence. Their attitudes both towards the course and towards the TSC approach were increasingly positive.

A second trend that can be observed from our data is that the progress described in the previous section was not always smooth and linear. By the middle of the course the students were equivocal in their opinions on thinking development. Some of the ideas presented in the course did not fit in with their prior knowledge. This caused concern, doubt, and even dilemmas about issues raised in the course. Feelings of bewilderment and even objections were common in the reports. Students were aware of difficulties and pointed out drawbacks and dilemmas associated with the TSC method.

By the end of the course, feelings of concern and bewilderment disappeared from students' reports. It seems that at this stage, they had acquired a new level of knowledge. Each student related to a wide range of aspects and used correct professional terminology. They made strong references to practical aspects of teaching by the TSC method (as opposed to the general theoretical ideas expressed at the beginning of the course). The concerns described at the middle of the semester gave way to optimistic feelings of improved competence and empowerment, expressing the need to continue using the TSC method in the future.

The knowledge acquired by students during the course refers to Shulman's category of pedagogical content knowledge (Shulman, 1986). This category includes knowledge about instruction of specific subjects. The knowledge is influenced by general pedagogical knowledge and by specific matter knowledge. In our case, it refers to procedural and metacognitive knowledge of thinking (Zohar, 1999).

In summary, students' development during the course can be characterized by the following three stages:

1. A stage of cognitive balance at the beginning of the course. At this stage students expressed intuitive ideas that were unambiguous but limited.
2. A stage of cognitive dissonance at the middle of the course. At this stage students may have known more about the issues taught in the course, but they also experienced confusion, lack of confidence, vagueness and mixed attitude towards the course.
3. A stage of restored cognitive balance at the end of the course, this time at a more advanced cognitive level.

The changes occurred in two domains:

1. The cognitive domain. Starting with simple, general ideas, which were then made more complex, accompanied by feelings of concern and bewilderment. Finally more knowledge was assimilated and the cognitive balance restored.
2. The affective domain. Starting with positive attitudes which were later distorted, expressing concerns and difficulties. Finally, positive attitudes were restored again together with a feeling of empowerment and ability to succeed in the challenging new instructional way.

Implications and Limitations

Implementation of higher order thinking in schools is nowadays a major goal for many educators. Finding adequate methods to prepare teachers for the complicated task of teaching higher order thinking is a considerable challenge. The major implications of this chapter are that it is possible to introduce this issue into teacher preparation programs and to evaluate its effect.

Our findings show that the course was effective in inducing changes in students' thinking and practice. However, we think that what we did has two limitations, which are constraints of the educational setting in which we work. We agree with our students' critique that the course was too short, as well as with their comments about the need to integrate higher order thinking into their science courses. We believe that the *Thinking in Science* course should have been longer to allow students to consolidate their new knowledge and ways of practice. Previous research conducted with inservice teachers using the TSC method has shown that continuous support from a university team during implementation of the method was a vital role to its success (Weinberger, 1992; Zohar, no date). We also believe that, according to the infusion approach, instruction of higher order thinking in teachers' colleges should take place in disciplinary science courses, not only in one special pedagogical course. However, these two desires could not have been accomplished in the educational setting where we work. Since similar practical constraints are prevalent in many educational institutions, it is especially important to realize that even a course of one semester may induce some change towards the goal of developing pupils' higher order thinking.

REFERENCES

Adams, P. E., & Krockover, G. H. (1997). Beginning science teacher cognition and its origins in the preservice secondary science teacher program. *Journal of Research in Science Teaching, 34*(6), 633-654.

Arter, J., & Spandel, V. (1992). Using portfolios of student work in instruction and assessment. *Educational Measurement: Issues and Practice, 11*(1), 36-44.

Bransky, J., Hadass, R., & Lubezky, A. (1992). Reasoning fallacies in preservice elementary school teachers. *Research in Science & Technological Education, 10*(1), 83-92.

Brown, A. L., & Campione, J. C. (1994). Guided discovery in a community of learners. In K. McGilly (Ed.), *Classroom lessons: Integrating cognitive theory and classroom practice* (pp. 229-272). Cambridge, MA: The MIT Press.

Brownell, G., Jadallah, E., & Brownell, N. (1993). Formal reasoning ability in preservice elementary education students: Matched to the technology education task at hand. *Journal of Research on Computing in Education, 25*(4), 439-446.

Bruer, J. T. (1993). *Schools for thought*. Cambridge, MA: The MIT Press.

Casey, M. B., & Howson, P. (1993). Educating preservice students based on a problem-centered approach to teaching. *Journal of Teacher Education, 44*(5), 361-369.

Clark, C. M., & Peterson, P. L. (1986). Teachers' thought processes. In M.C. Wittrock (Ed.), *Handbook of Research on Teaching* (3rd ed.) (pp. 225-296). New York: Macmillan.

DeRose, J. V., Lockard, J. D., & Paldy, L. G. (1979). The teacher is the key: A report on three NSF studies. *The Science Teacher, 46*(4), 31-37.

Donnellan, K. (1982). *NSTA elementary teacher survey on preservice preparation of teachers of science at the elementary, middle, and junior high school levels.* Washington, DC: National Science Teachers Association.

Dossey, J.A., Mullis, I. V. S., Lindquist, M. M., & Chamber, D. L. (1988). *The mathematics report card: Are we measuring up?* Princeton, NJ: Educational Testing Service.

Dressler, T., & Levinger-Dressler M. (1997). *Scientific and technologic education in elementary school.* Lamda, Tel Aviv: Tel Aviv University. (in Hebrew)

Duschl, R. A. (1990). *Restructuring science education: The importance of theories and their development.* New York: Teacher College Press.

Ennis, R. H. (1989). Critical thinking and subject specificity: Clarification and needed research. *Educational Researcher, 18*(3), 4-10.

Feurstein, R. (1991). Intervention programs for retarded performers: Goals, means, and expected outcomes. In L. Idol & F. L. Beau (Eds.), *Educational values and cognitive instruction: Implications for reform* (pp. 139-175). Hillsdale, NJ: Erlbaum.

Feurstein, R., Rand, Y., Hoffman, M. B., & Miller, R. (1980). *Instrumental enrichment: An intervention program for cognitive modifiability.* Baltimore: University Park Press.

Friedler, Y., & Tamir, P. (1984). Teaching and learning in the laboratory in high school biology classes in Israel. *Research in Science Education, 14*, 89-96.

Grossman, P. L., Wilson, S. M., & Shulman, L. S. (1989). Teachers of substance: Subject matter knowledge for teaching. In R. Maynard (Ed.), *Knowledge base for the beginning teacher* (pp. 23-36). Oxford: Pergamon Press.

Harary, H. (1992). *Report of the higher committee for science and technology education – "Tomorrow 98".* Jerusalem, Israel: Israeli Government. (in Hebrew)

Hewson, P.W., Kerby, H.W., & Cook, P.A. (1995). Determining the conceptions of teaching science held by experienced high school science teachers. *Journal of Research in Science Teaching, 32*(5),503-520.

Krombey, J. D. (1991, February). *Toward establishing relationships between essential and higher order teaching skill.* Paper presented at the annual meeting of the Eastern Educational Research Association, Boston, MA.

Kuhn, D., Garcia-Mila, M., Zohar, A., & Anderson, C. (1995). *Strategies and knowledge acquisition. Monographs of the society for research in child development* (Vol. 60(4)). Chicago: University of Chicago Press.

Lipman, M. (1985). Thinking skills fostered by philosophy for children. In J.W. Segal, S.F. Chipman & R. Glaser, (Eds.). *Thinking and learning skills.* (Vol. 1) (pp. 83-108), Hillsdale, NJ: Erlbaum.

Mendelowitz, R. (1996). *Teaching biology in Israeli high school, at the middle of the 90s, emphasizing inquiry.* Unpublished study. (in Hebrew)

Miles, M. B., & Huberman, M. (1994). *An expanded sourcebook: Qualitative data analysis.* Newbury Park, CA: Sage Publications.

Ministry of Education and Culture. (1954). *Nature and agriculture curriculum.* Jerusalem, Israel: Author. (in Hebrew)

Ministry of Education, Culture and Sport. (1996). *Science and technology for junior high.* Jerusalem, Israel: Author. (in Hebrew)

Mullis, I. V. S., & Jenkins, L. B. (1988). *The science report card: Elements of risk and recovery.* Princeton, NJ: Educational Testing Service.

Mullis, I. V.S., & Jenkins, L.B. (1990). *The reading report card, 1971-1988.* Washington D.C.: Office of Educational Research and Improvement, US Department of Education.

Perkins, D. (1992). *Smart schools: From training memories to training minds.* New York: The Free Press.

Resnick, L. (1987). *Education and learning to think.* Washington, D.C.: National Academy Press.
Schön, D. A. (1983). *The reflective practitioner: How professionals think.* London: Temple Smith.
Schön, D. A. (1987). *Educating the reflective practitioner.* New York: Basic Books.
Sesow, F. W. (1991). *Improving the critical thinking ability of preservice social studies teachers* (Report No. 143). Nebraska: U.S. (ERIC Document Reproduction Service, No. ED 356 997)
Shayer, M., & Adey, P. S. (1992a). Accelerating the development of formal thinking in middle and high school students 2: Post project effects on science achievement. *Journal of Research in Science Teaching, 29*(1), 81-92.
Shayer, M., & Adey, P. S. (1992b). Accelerating the development of formal thinking in middle and high school students 3: Testing the performance of effects. *Journal of Research in Science Teaching, 29*(10), 1101-1115.
Shulman, L. S. (1986). Those who understand: Knowledge growth in teaching. *Educational Researcher, 15*(2), 4-14.
Siegler, R. S., & Jenkins, R. (1989). *How children discover new strategies.* Hillsdale, NJ: Erlbaum.
Tamir, P. (1985). The Israeli "Bagrut" examination in biology revisited. *Journal of Research in Science Teaching, 22*(1), 31-40.
Weinberger, Y. (1992). *Fostering critical thinking in biology instruction.* Unpublished Masters Thesis, Tel Aviv University, Tel Aviv, Israel. (in Hebrew, with an English summary).
Yakobi, D., & Sharan, S. (1985). Teacher beliefs and practices: The discipline carries the message. *Journal of Education for Teaching, 11* (2), 187-199.
Zohar, A. (1996). Transfer and retention of reasoning strategies taught in biological contexts. *Research in Science and Technological Education*, 14 (2), 205-219.
Zohar, A. (1999). Teachers' metacognitive knowledge and instruction of higher order thinking. *Teaching and Teacher Education, 15*, 413-429.
Zohar, A. (no date). *Thinking teachers: conflicts between teachers' cognition and higher order thinking in science classrooms.* Manuscript submitted for publication.
Zohar, A., Schwartzer, N., & Tamir, P. (1998). Assessing the cognitive demands required of students in class discourse, homework assignments and tests. *International Journal of Science Education, 20*, (7), 769-782.
Zohar, A., & Weinberger, Y.(1995).*Thinking in science.* Jerusalem: Science Education Center, The Hebrew University of Jerusalem (in Hebrew).
Zohar, A., Weinberger, Y., & Tamir, P. (1994). The effect of the biology critical thinking project on the development of critical thinking. *Journal of Research in Science Teaching, 31*(2), 183-196.

Chapter 7

Secondary Science Student Teaching Assessment Model

A United States and United Kingdom Collaborative

Kate A. Baird[1], Marilyn M. Brodie[2], Stuart C. Bevins[2], and Pamela G. Christol[3]

[1]Designs on Learning, U.S.A.: [2]Sheffield Hallam University, U.K.: [3]National Aeronautics and Space Administration/Aerospace Education Services Program, U.S.A.

Abstract: We studied current practices in teacher preparation and student teaching assessment at two universities--Sheffield Hallam University in the United Kingdom and Oklahoma State University in the United States--and identified current practices through university documents and participant interviews in an effort to develop a recommended model for student teacher assessment. The recommended model is based on theory as identified through national standards, literature review, and interviews, and incorporates the practical through participant as well as professional organization recommendations.

Over recent years there have been a number of attempts to change the way in which teachers are prepared. The foundation for many of these reforms is the perceived dichotomy between theory and practice. This study looks at how two secondary science teacher preparation programs--Oklahoma State University (OSU) in the United States (US), and Sheffield Hallam University (SHU) in the United Kingdom (UK)--balance theory and practice within the student teaching component of their programs. The theoretical aspect of each university program was identified through examining their policy and course documents. Recent participants of the programs--student teachers, cooperating teachers, and university supervisors--were interviewed concerning strengths and weaknesses of the current programs to understand their practical aspects. A unified model was then constructed to promote a balance of theory and practice in student teacher assessment. It is important to note here that it is the similarities between the K-12 national science education standards for both countries that made this modeling possible.

S.K. Abell (ed.), Science Teacher Education, 121–140.

DESCRIPTION OF THE STUDY

The design of this study is both qualitative and quantitative in nature. We describe science education in both countries, their national standards, the general characteristics of each institution, and the route to certification/diploma at each. In addition to this general information, we analyze the assessment methods used in student teaching for commonalties. To begin the study, a review of the student teaching process at both universities was completed and copies of all forms used to evaluate student teachers were collected. Next interviews with student teachers, cooperating teachers, and university supervisors were conducted. We used the combined information to design a best practice model reflecting the balance of theory and practice in the standards, current assessment methods, and participant suggestions.

Assessment criteria related to student teacher evaluation were organized into categories related to personal and professional competencies, and descriptors related to these competencies were compared with the US National Science Education Standards (National Research Council, 1996), specifically the Science Teaching Standards (pp. 27-53) and the Professional Development Standards (pp. 55-73) and with the Department for Employment and Education (DfEE) standards in the UK (Department for Education and Science, 1992).

ISSUES IN TEACHER PREPARATION

University-Based Environment

Teacher education has received recurrent investigation, and although many studies have been published concerning issues related to teacher education, enduring problems in the field still exist. Conant's 1963 study reiterates the lack of common course requirements available among colleges (in Lanier & Little, 1986). In the US, the Holmes Group has proposed one model of teacher preparation while the National Council of Accreditation for Teacher Education (NCATE) and the content areas through their professional organizations such as National Science Teachers Association (NSTA) are proposing other models. Some agreement exists among the varying interests, however, no clear model of best practices exists.

Typically in 4-year colleges and universities in the US, nearly three quarters of the course work for prospective secondary science teachers is organized into three areas: general education, subject matter concentration, and pedagogical study. Beyond these three areas, initial preparation expectations lack common substance and sometimes even vary within the same institution. Considerably more courses are taken outside professional education than are taken within it (Lucas, 1997). The confusing disparity of requirements and electives have led to a lack of common intellectual experience and in some cases inconsistently or ill-prepared teachers.

In the past, there has been little agreement related to a common body of knowledge that all schoolteachers should possess (Lanier & Little, 1986). In 1987, a National Board for Professional Teaching Standards was created in the US to improve the status of the teaching profession by generating standards for teacher knowledge and actions. Accompanying the standards are assessment procedures that involve portfolios, student interviews, simulated performance, and on-site observations. National certification is granted in specific areas to applicants who have proven achievement of the standards. National certification could provide assurance of high quality and enable teachers to seek employment in all fifty states without requiring state-by-state licensing. Currently a standards-based movement for beginning teachers, led by the Interstate New Teachers Assessment and Support Consortium (INTASC) (1992) has begun to drive teacher preparation and licensure across the US.

The development of standards for global/international education is currently underway through the efforts of the Council of Chief State School Officers in the US, the Elliot School of International Affairs of The George Washington University, and the American Forum for Global Education. The global standards offer guidelines for teaching youth throughout the world. This includes knowledge of content areas, skill competencies, and attitudes (Haakenson, 1994). Global/ international education would enable qualified teachers an opportunity to understand learning as an international activity, which will enhance the likelihood that they will imbue their students with a global perspective.

In the UK, to meet the demands of the National Curriculum, the Department for Employment and Education (DfEE) published Circular 9/92 (Department for Education and Science, 1992), which established a new Secondary Initial Teacher Training (ITT) program. The Circular introduced new criteria and procedures for England and Wales for the accreditation of courses in ITT. Keys to the new program were identified as:

- schools should play a part in ITT as full partners with Higher Education Institutions
- the accreditation criteria for ITT courses should require Higher Education Institutions, schools and students to focus on the competencies of teaching
- Institutions rather than individual courses, should be accredited for ITT.

It is the first point that we examine in more detail below.

School-Based Environment

The view that theory is over emphasized and perhaps unnecessary (Lawlor, 1990; O'Hear, 1988) has fueled the drive to a more school-based approach to Initial Teacher Training (ITT). What is interesting is that research into student teachers' perceptions of the theory-practice debate (Lanier & Little, 1986) shows that they too are disillusioned. The disillusionment appears to result from the lack of linkages between experiences within the classroom and university based components of their teacher preparation programs (Calderhead & Shorrock, 1997). Students find it difficult to build on academic experiences while dealing with the complex, real world events of the classroom. This issue highlights a lack of collaboration between the host school and the university.

Calderhead and Shorrock (1997) asked the question, "Why do practicing teachers themselves often not rate their professional training highly?" (p.10). They suggested that teaching has a culture that emphasizes the importance of personal and practical experience and attaches less significance to academic and professional qualifications. This claim is grounded in the action oriented nature of teaching. This culture is best associated with teachers' work rather than reflection or reading. While it is fair to say that teacher effectiveness is often related to years of experience, there is also a danger in undermining teacher professionalism if we do not associate reflection and reading equally with the classroom practice element of the profession. Experience should be seen as a vehicle for the teacher to mature into a master teacher.

McNally, Cope, Inglis, and Stronach (1994) suggested that classroom practice enables student teachers to develop emotional links with teachers and children. This sociocultural aspect of ITT provides a more memorable account than the less dynamic experiences of many university-based experiences. These experiences may explain the lack of enthusiasm when teachers evaluate their ITT programs. Their school experience may be more vivid and easier to recall. In a number of detailed and descriptive case studies of secondary school student teachers, Grossman (1992) found that a great deal of student teachers' planning and teaching was directly influenced by the university component of their ITT's. In contrast to Calderhead and Shorrock's statement, Grossman found that students acknowledged the value of the university as they taught particular lessons. Although the students in Grossman's study were able to reconcile classroom and university based components, it appears that they found "real" experiences of the classroom easier to emphasize.

Considerable research on teacher development (Barba & Rubba 1993; Carter, Cushing, Stein, and Berliner, 1988; Carter, Sabers, Cushing,

Pinnegar, and Berliner, 1987; Sanders, Borko, and Lockard, 1993) has direct implications for student teaching. Berliner's (1988) model of teacher development explains student teaching as rational, inflexible, and needing full concentration. His study indicated that student teachers attempt to develop by understanding rules of teaching. Unfortunately, unless the initial anxieties of teaching can be alleviated quickly, the move from rational, highly planned routines to more creative, personal practices may be delayed. Kagan's (1992) description of student teacher development includes an increase in metacognition and a move from the self to instructional design of pupil learning (Kwo, 1994). This increase in metacognition and move to a more personalized practice emphasizes the importance of the experiential context of student teaching. However, theories of metacognition and reflection (Schön, 1983; 1987) are possibly better examined within the university component and then brought into the classroom. This research identifies a salient implication to teacher preparation.

Teacher educators need to develop an epistemological stance that, while addressing the practical experience of student teaching, enhances a collaborative framework that addresses both theory and practice. The partnership between host school and the university is a fundamental construct for the collaborative framework; it requires clear definition with regard to the roles of each partner, roles that must be nurtured and maintained through effective communication and mutual respect.

Mentoring by experienced teachers has a critical role to play within student teacher development; this role also requires support to the mentor as well as time to enable the mentor to provide advice and support for the student. Mentors must be empowered, thereby strengthening their professionalism and improving their role as student mentor. Lawlor (1990) and O'Hear (1988) emphasized the issue of adding the mentor to the balance of the relationship between the university and the school, and cautioned that to abolish the university role would place student teachers within a context of apprenticeship learning. In a recent study of teachers' perceptions of mentoring (Jones, Reid, and Bevins, 1997) there was little evidence to suggest that the apprenticeship model was preferred to the academic input model by mentors. Mentors confirmed their regret that they were unfamiliar with current educational theory. Moreover their argument for a continuing role for universities within ITT is a rational one as suggested by Furlong and Smith (1996). Ongoing mentor development activities must not be allowed to lapse into only a discussion of years of experience. The award of Qualified Teacher Status (QTS) /state licensure should not be enough to establish a teacher as a quality mentor. There must be a forum that promotes concepts of mentor quality that is validated within higher education programs and by nationally recognized criteria.

Teacher involvement in teacher preparation is essential. Teachers need a voice in the establishment of new teacher entry standards and courses. The introduction of a General Teaching Council and other professional organizations may go some way to enhancing teacher participation (Barba, 1993). Practical experience is central to ITT and mentor teachers are the keys. The role of student teaching in the professional development of teachers is not simply one of teaching practice, but something much larger. Teacher preparation both in the university and school settings will then allow the diversity of activities and experiences to be assimilated by the evolving student teacher.

SCIENCE TEACHER EDUCATION

In the United States

Science teacher preparation in the US is facing a period of many changes. Without a national curriculum or certification, teacher preparation continues to be the purview of the individual states and institutions of higher education. Moves do exist to bring consistency to teacher preparation through professional education organizations such as the National Science Teachers Association (NSTA) and the National Council for the Accreditation of Teacher Education (NCATE). However, many states still expect teachers to find their own model of teaching excellence. Since the introduction of the INTASC standards (1992) and the National Science Education Standards (NSES), universities and states have begun to rethink teacher preparation. Several universities and states, such as New York State, have moved to mandated master's degrees before certification, although this is not the norm in the US. Most teacher preparation programs are based on 4-year programs that combine science content with educational psychology and classroom management. Students spend the majority of their time in content based math and science courses. These courses are taught in a traditional lecture and lab format. The students have little opportunity to experience true inquiry and independent or small group instruction. Clearly the role models for instruction are heavily based in the science content.

Program at Oklahoma State University

Education majors are required to complete 40 semester hours of course work in education, and maintain good academic standing, which is defined as a minimum grade point average of 2.5, (this equates to a C+ or about 75%). The last semester of their senior year, students are required to complete and pass 12 weeks of a student teaching practicum in which they are required to gradually take over the responsibilities of teaching until two consecutive weeks of full-time teaching are completed. During the 12 weeks of student teaching, the activities are monitored and observed by a cooperating teacher within the school and a supervisor representing the university. Table 1 provides a comparison between the typical US teacher education program and the program at OSU. Note that OSU is on a semester system. This means that students typically spend one hour of in-class instruction for 15 weeks to equal one semester credit hour.

Table 1. Typical US Science Teacher Preparation Experiences

Experience	Typical Semester Hours	OSU* Semester Hours
Math and Science Instruction	50-75	40
Educational Psychology	6-9	(included above)
Classroom Management (methods/student teaching)	20-24	10-16 weeks

*OSU=Oklahoma State University

The student teacher is placed in a school after a comprehensive professional review of his/her record indicates that academic performance, professional competence, and personal character qualify him/her for student teaching. Cooperating teachers are required to have a minimum of three years teaching experience. Most cooperating teachers self-select to work with student teachers. However, due to minimal financial rewards, some administrators take control of the process. Once a match as been made, the student is encouraged, but not mandated, to make early contact with the cooperating teacher. The university distributes a pre-printed "teachers handbook" via the student teacher upon his/her arrival at the school. The university's next contact with the school is during the first of three or four site visits. The school then receives a final evaluation form at the time of completion of the placement.

The cooperating teacher and university faculty do not experience any pre-placement activities or training. Furthermore, a university professor may or may not supervise any given student. Much of this supervision is assigned to graduate assistants. The cooperating teacher completes a final evaluation that may or may not be used to inform the grade (pass/fail). Cooperating teachers have had only minor input into the actual structure of the evaluation form. Similarly, graduate teaching assistants use forms created by the course instructor for whom they work. These may or may not be referenced to state or national standards or competencies.

In the United Kingdom

The UK education system is currently undergoing major reforms in a bid to raise standards. With the low enrollment in science courses at post compulsory level, science education has been targeted as a priority subject in both learning and teaching. There will be cash incentives for graduates considering teaching courses in the key shortage subjects (science and mathematics). In addition to this, a new scheme, The Graduate Teacher Program, was introduced early in 1999. Six hundred trainee teachers, who will train 'on-the-job,' will be placed in vacancies arising in schools within the subjects of science and mathematics. A new national curriculum details the science content which student teachers must learn together with Information and Communication Technology (ITC) skills in all subjects. These changes are an attempt by the government to inject a high level of quality within education in the UK, particularly in classroom teaching.

Program at Sheffield Hallam University

All students in the secondary schools program undertake the Professional Year that is the final year of 2 and 3-year undergraduate routes or the 1-year postgraduate route. The Professional Year includes three components: school experience, applied education, and subject application (see Table 2). Each component is assessed with competency in practical teaching twice during the year. Mentors in the schools assign grades on a pass/fail basis by using the Profile of Developing Professional Competence (see Table 3).

Table 2. Typical UK Teacher Preparation Experiences

Experience	Typical Hours per Semester
Math and Science Instruction	100
Education Psychology	36
Classroom Management (methods/student teaching)	included in Math and Science
Professional Year	12 weeks University 24 weeks School

Note: SHU measures total hours per semester. Thus 36 hours per semester would be just over 2 semester credit hours in the OSU system.

The program was developed to reflect the changing needs of the schools into which students go; each year of the program has its own specific needs and rationale all feeding into the professional Year. Four routes to program completion exist: 3-year BSc Hons + Qualified Teacher Status (QTS), 3-year BSc + QTS, 2-year BSc + Hons + QTS, and 2-year BSc + QTS. The program meets the needs of both undergraduates and postgraduates and is integrated by the common Professional Year, which contributes to the QTS embedded in the program.

Table 3. Profile of Developing Professional Competence

Area	Focus
Curriculum Knowledge	• Knowledge of subject and curriculum
Planning	• Planning for individual sessions and program of work, selecting appropriate materials
Pupil Assessment	• Monitoring, assessing, recording, interpreting learner performance
Classroom Organization	• Organizing and managing learning environment
Teaching Skills	• Implementing and follow through of programs
Managing Pupils	• Establishing good relationships and organizing and managing learners
Self-Evaluation	• Evaluating self on teaching and implementing modifications as appropriate
Professional Skills	• Understanding teaching role and responsibilities
Personal Contribution	• Setting and maintaining standards of professional behavior

Since 1994, a large number of schools have worked in partnership with Sheffield Hallam University (SHU) to provide initial training for all secondary students in Professional Year programs. The school provides not only the site of the placement but also the mentor. The mentor is a cooperating teacher, frequently the science chair, who sets up the learning situation and aids in the development of the competencies. The mentor is trained by the university and observed by the university supervisor at least once during the process. The mentor is the ultimate authority over the student's placement and the inherent activities within that placement. Students work in two schools during their student teaching. Students from the SHU program may be placed in schools throughout the UK. The schools are both public and private and can be rural or urban in location. SHU is also actively involved with schools in London. The diversity of possible placements increases student opportunities but also increases the difficulties in maintaining strong partner-like interactions.

Table 4 summarizes the size of the teacher preparation programs at SHU and OSU as well as their student teaching sites.

Table 4. Comparison of Teacher Education at Two Universities

Country	Institution	Size of Teacher Preparation Program	Size and Type of Student Teaching Sites
US	Oklahoma State	1500 education students	80-3000 students; rural/urban
UK	Sheffield Hallam	1500 students in Initial Teacher Training	500-1500 students; rural/urban

EMERGENCE OF PROFESSIONAL COMPETENCE

Overview of US standards

Work has been done by several professional organizations in the US in an effort to address concerns in teacher preparation in light of the NSES. One such effort is the Certification and Accreditation in Science Teacher Education (CASE) project (National Science Teachers Association [NSTA] and the Association for the Education of Teachers in Science [AETS], 1998). The standards have been developed by the NSTA and the AETS with a goal of excellence in science teacher education. Table 5 summarizes the key areas addressed by the standards. The driving philosophies behind these standards are those of inquiry-based instruction for all learners.

Table 5. CASE Standards Summary

Area	Focus
Science Content	• Concepts, principals, and relationships of science
Nature of Science	• Creation of scientific knowledge based on values, beliefs, and assumptions
Teaching through Inquiry	• Teaching the processes of investigations such as problem solving, data collection, theory analysis, and communication of findings
Context of Science	• Framework of science as a human endeavor and everyday process
Pedagogy	• Development of communities of diverse learners of science
Science Curriculum	• Planning of instructional goals, plans, and materials with appropriate pedagogy
Social Context of Science Teaching	• Community in which learning is taking place
Assessment	• Authentic and equitable assessment strategies
Environment of Learning	• Location where learning is taking place—safe, healthful, and supportive of learning
Professional Practice	• Community of teaching professionals

From: NSTA and AETS, 1998

Overview of UK standards

Fundamental to each of the professional year routes in the UK is a belief that, as a result of their studies, students will be prepared to commence their careers as competent, confident, well organized, and thoughtful beginning teachers. The routes offered are seen as but the first stage of professional development, providing a basis for induction program in schools and further in-service career development. Beginning teachers need to have appropriate attitudes and skills that will prepare them to meet the demands and challenges presented by the national curriculum and its accompanying assessment procedures, changes in the management of schools, and economic and technological developments.

To assist with student assessment, the Profile of Developing Professional Competence was developed. It provides a framework to regulate discussion between student and mentor based on evidence and shared experience. Table 3 summarizes the categories and content/skills addressed by the profile.

Standards Comparison

A direct comparison of the science content and process standards for K-12 students in both countries showed a complete agreement. Topics, vocabulary and skills are parallel in both countries. Direct comparison of teacher preparation standards was difficult to assess. The CASE and INTASC standards were not in use at OSU at the time of the study. The Competence Profile in the UK suggests curriculum opportunity, but is not binding. Thus the general intent of both programs is similar enough to allow for direct comparison and application within specific programs of teacher preparation.

Student Teacher Assessment Comparison

The assessment systems for student teachers utilized in both the US and UK institutions are similar in style, provider, and frequency (see Table 6). University-provided assessment forms are also the norm. However, in the UK, a key difference in the responsibility of the university supervisor requires clarification. In the UK, the university supervisor's role is to assess the mentor (classroom teacher) instead of the student.

DESCRIPTION OF DATA COLLECTION

In April 1996, researchers in the US and the UK agreed to form small focus groups of representative students in each country. The purpose of these groups was to lead to the development of a set of interview questions. The first round of interviews led the researchers to adjust the interview strategy. The initial interviews showed that the interviewees not only reflected on the frequency of assessment, the types of assessment, and the variety of assessors, but also the roles that each participant played throughout the placement. The final design was for a series of unstructured

Table 6. Characteristics of Student Teacher Assessment Systems in Two Institutions

Subject of Assessment	Institution	Style of Assessment	Provider of Assessment	Frequency	Forms Used?
Student	OSU*	Observation	University supervisor	2-3 events	Yes
			Classroom teacher	Daily feedback	Yes
		Interview	Classroom teacher/ administrator	Single exit event	No
	SHU*	Observation	Classroom teacher	Daily feedback or single exit event	Yes
Classroom Teacher	OSU	Observation	Classroom teacher	Varied	Univ. provided
			University supervisor	2-3 in 10-12 weeks	Varied
	SHU	Observation	Staff member/Lab technician	Daily	Yes
University Supervisor	OSU	Observation	University supervisor	Every other week	Varied
		Feedback	Classroom Teacher	Weekly	No
	SHU	Assesses classroom teacher, not the student		1-2 per placement	No

*OSU=Oklahoma State University; SHU=Sheffield Hallam University

interviews that required the student teachers, cooperating teachers, and university supervisors to reflect on the placements that they had just completed.

The interviews were conducted in person at the school or through electronic communication from the school. This caused some variation in interview length due in part to stresses from the activities and teaching of the day. The interviews took place in two or more sessions with no session taking more than one hour. Following the final interview, the interviewees were asked to comment on the interviewer's notes. This was done to decrease the likelihood of comments being taken out of context by the interviewer.

Between 1997 and 1998 approximately 15% of the students completing certification in secondary science education at the partnership universities participated in these interviews. The number of cooperating teachers interviewed was smaller, with approximately 10% of the cooperating teachers participating. Some of the teachers worked with several student teachers during the study. Teachers interviewed during the first placement were not asked to participate subsequently. The monitoring and assessment processes followed by the universities did not change during this time; therefore, cooperating teachers and university supervisors were not re-interviewed. Twenty five percent of the university supervisors were interviewed. The other 75% were involved in the study as interviewers and researchers.

Teachers and university personnel from both countries were asked to discuss suggestions for strengthening the assessment process in which they had recently been engaged. Improvement themes surrounding the student teaching process were primarily limited to the process of student evaluation (Table 7). References by classroom teachers to opportunities for interaction with their peers and by university teachers to train school personnel address different goals felt by study participants. A key dissimilarity in student grading was expressed by classroom teachers, who desired to assign student

Table 7. Process Evaluation Characteristics

Respondent	Process Improvement Suggestions												Assessment Form Improvement Suggestions		
	Increase Observation Event	Increase Observation Duration	Increase Team Interactions	Standardize Evaluation Rubrics	Link Letter Grades to Narraative Evaluations	Assign Grades Based on Common Expectations	Focus on Clarity of Written Forms	Increase Interactions Between University and School	Allow Classroom Teachers to Assign Grades	Grade Student in Team Performance Review	Provide Mentors Opportunity to Watch/Work with Peers	Train School Personnel	Resign Forms for Specialized Content and Grade Levels	Make Forms Parallel to Those for First Year Evaluation	Improve Readability, Directions, Grading Scales
Students	✓	✓	✓	✓	✓	✓	✓						✓	✓	
Classroom Teachers								✓	✓		✓				✓
University Teachers	✓	✓								✓		✓			

teacher grades themselves, vs. the university teacher, who desired to assign grades in a team performance review. Additionally, both students and classroom teachers expressed needs for improvement in form design and content.

RECOMMENDATIONS FOR A UNIFIED MODEL

A unified model for monitoring and assessing secondary science student teachers is needed to complement the currently evolving science education reforms. The model defined by this study calls for assessment practices based on narrative information gained from observations, portfolios, and joint interactions between the classroom mentor teacher, student teacher, and university supervisor. This team would jointly critique the success of the experience, with each team member contributing key information. Assessment of the student teacher would focus on the concepts identified in the roles of the student teacher, classroom mentor, and the university teacher as defined in Tables 8, 9, and 10. Additionally, standard assessment forms would evolve to allow ease of documentation and communication of assessment findings. Although the recommendation's prime objective is to improve student teacher assessment practices, it also defines responsibilities of the university supervisor to both the classroom mentor/school and the student teacher, the classroom mentor to the student teacher and the university, and the student teacher to the classroom and the university/professional environment.

Following each table are selected quotes from student teachers, mentor teachers, and university supervisors that provide support for the model. The interview questions did not focus directly on the proposed model, yet many of the anecdotal comments supported components of the model. Many of these comments are an indictment of the current process, while others are evidence of parts of the current process that seem to be working well. It is our intent to give voice to those most directly involved in this study-- but the comments included are not an exhaustive listing. The speakers have not been identified in any way to protect their privacy and prevent any inadvertent discrimination due to less complimentary comments. However after each quote the type of participant is identified as student teacher (S), cooperating teacher (T) or university supervisor (U).

Table 8. Recommended Model for Student Teacher Roles

Roles	Description
Collaborator	• Incorporate input from mentor and university supervisor, work with school staff, work with parents and pupils • Provide frequent and regular review of progress
Learner	• Be open to instruction from all, select appropriate practice, begin to think critically
Pupil Supporter	• Identify needs of individuals and adjust to those needs
Active Listener	• Show willingness and awareness of the teaching environment
Meaning Maker	• Provide pupils with subject understanding • Sort out daily experiences
Linker	• Share in transfer of current pedagogy and practice between university and school
Reflective Practitioner	• Use the information gained through listening
Problem Solver	• Identify situation, describe full image, discern problem areas, generate divergent solutions, evaluate possible solutions, act responsibly
Proactive Learner	• Seek solutions without being prompted

Table 8 describes the recommended roles for the student teacher. Comments from the interviews in support of these student teacher roles include:

- Learner: *I do like receiving a copy of the supervising teacher's comments and observations so that I can become aware of weak areas and strong areas.* (S)
- Active Listener: *Being evaluated by the woman for whom I student taught meant more to me because she was actually a teacher, and she saw me every day.* (S) *I was overwhelmed trying to learn about all 100 students in less than eight weeks. If it had not been for my cooperating teacher I never would have made it.* (S)
- Linker: *Student teaching was a constant learning experience and I found myself being a teacher and student interchangeably. This was true not only with the students but also with the university supervisor.* (S) *I occasionally act as a source of information for my Cooperating Teacher.*(S) *I have brought my teacher up to date on some of the latest developments in the field of education.* (S) *I have special skills in geology and was given the chance to share these with not only the students, but also the classroom and university supervisors.* (S)
- Reflective Practitioner: *The journal was good for self evaluation.* (S) *Informal evaluation was the most helpful. They wrote down what they saw. The teacher then told me what I did that was good, suggestions for other methods, etc.* (S)

- Proactive Learner: *I suggest the university offer a course dealing with discipline and parent/teacher interactions.* (T)

Table 9. Recommended Model for Cooperating Classroom Teacher Roles

Role	Description
Daily Facilitator	• Provide access to teacher handbook, time table, initiation into whole school
Role Model	• Provide access to all of the daily routines of a teacher as well as culture of the school
Pedagogical Facilitator	• Demonstrate appropriate teaching and learning strategies
Collaborator	• Seek and incorporate input from student teacher and university supervisor, work with staff, work with parents and pupils
Professional Education Model	• Introduce student teacher to mutual respect and understanding, describe strategies for agreeing to disagree
Resource Person	• Share reference and curriculum materials, experiences, introduce professional organizations • Share strategies for grant writing and materials collection
Linker	• Share current pedagogy and best practices between university and pupil
Professional Educator	• Be aware of and follow best practice as determined by professional experiences and organizations • Actively seek and support pupil learning
Goal Setter	• Establish boundary and targets, help develop a plan of attack, and provide opportunity for guided practice while transferring responsibility for learning to the student
Counselor	• Listen actively, decrease stressful situations, identify land mines • Provide frequent and regular review of progress
Active Listener	• Show willingness and awareness of the teaching environment

Table 9 describes the suggested roles for the cooperating teacher. [illegible] Classroom teacher roles, the following comments were observed.

- Daily Facilitator: *He also suggests community resources that I was not aware of.* (S)
- Role Model: *He is an excellent model of an easy going and humorous teacher who also has the ability to crack down on problems as they arise.* (S)
- Professional Education Model: *Our relationship has been characterized by mutual respect and an easy ability to work together.* (S) *While we work together throughout the day, we switch between these roles freely and frequently.* (S) *I was really more of a co-teacher than a student teacher. My mentor and I freely exchanged roles within each lesson.* (S)

- Resource Person: *I work very hard to make student teaching meaningful to the student while allowing them to find their own style.* (T)
- Goal Setter: *My student teaching experience failed to meet my expectations because it simply unfolded without real planning, without discussion/ dialogue, without scrutiny from anyone.* (T) *I really didn't feel that there was any attempt to look critically at what I was doing and to help me improve in any meaningful way.*(T) *My particular evaluation from the supervising teacher was written observations only. I suggest combining this with a standard rubric for student teachers.* (T)
- Active Listener: *He has listened to my suggestions, asked questions, and has stepped aside to let me work.* (S)

Table 10. Recommended Model for University Supervisor Roles

Role	Description
Theoretical Liaison	• Research and communicate findings about science, science learning, and science teaching
Collaborator	• Link between student and school to transfer current pedagogy and practice • Provide frequent and regular review of progress
Resource Person	• Lifetime reference for curriculum materials, experiences, and perspective
Supporter	• Provide ideas about pedagogy, professional development, content and practice, and the developmental nature of teaching; act as a sounding board
Best Practice Modeler	• Prescribe the expectations for a successful placement
Active Listener	• Show willingness and awareness of the teaching environment
Counselor	• Be detached from school; provide a shoulder to cry on
Induction Mentor	• Educate future mentor into the structures and roles of the partnership (team)
Critical Coach	• Provide link to professional standards

Table 10 describes the recommended roles for university supervisors. Interview quotes regarding the university roles follow:

- Theoretical Liaison: *My methods classes were ok, but I really discovered what I should have been paying attention to after I student taught.* (T) *I wish I could have a follow-up class to go over all the questions and things that came up in my placements.* (S)
- Resource Person: *I don't know where she finds all those books and ideas. She [university supervisor] must have a complete library of every science book ever written.* (S)
- Supporter: *Some forms are helpful. They keep the information organized and let the student teachers know where the evaluation emphasis is.* (U)

- Best Practice Modeler: *I found team planning to be of the greatest help in my student teaching. The interaction regarding the possible teaching strategies was invaluable. I feel that I now understand some of the complexity in daily planning.* (S)
- Counselor: *I had some real problems with my school. . . I'm glad that I had someone outside that I could talk to.* (S)

CONCLUSION

Earlier in the paper we proposed that student teaching requires clear definition with regard to the roles of each partner that must be nurtured and maintained through effective communication and mutual respect. The interview quotes draw into critical relief the importance of communication, but also the importance of mutual respect among all members of the team. The evidence indicates that many students have had positive experiences in their student teaching, however the number who had grave concerns is alarmingly high. Areas of frequent concern include the interactions of the team members, the clarity of roles, and progress reporting.

Just as the US and UK educational systems have evolved similar processes for assessing student teachers, other areas in the world may also be able to use the proposed model. With the opening of curtains and breaking down of walls, additional opportunities exist to influence emerging educational systems. The unified model of student teaching assessment presented here can become a basis worldwide with the support of professional science education networks and practitioners. Teachers and students are citizens of the world and it is time that education takes an active role in smoothing the way for global communities of learners. After all, teachers of science in two highly industrialized countries have arrived at the same models of teacher preparation and assessment with very little interaction in differing time frames. Joint models, like the one presented here, are collaborative events that can be learning forums for all participants

AUTHOR NOTE

This study is part of an ongoing project conducted in partnership with Science International, SUNY Oneonta, and Sheffield Hallam University. Partial support came from local research funds from Oklahoma State University, The Centre for Science Education, and the Science International Trust.

REFERENCES

Barba, M. (1993). The truth about partnership. *Journal of Education for Teaching, 19*(3), 255-262.

Barba, R.H., & Rubba, P.A. (1993). Expert and novice Earth and space science teachers' declarative, procedural and structural knowledge. *International Journal of Science Education, 15*(3), 273-282.

Berliner, D.C. (1988, February). *The development of expertise in pedagogy.* Charles W. Hunt Memorial Lecture presented at the annual meeting of the American Association of Colleges for Teacher Education, New Orleans, LA.

Calderhead, J., & Shorrock, S.B. (1997). *Understanding teacher education: Case studies in the professional development of beginning teachers.* London: Falmer Press.

Carter, K., Cushing, K., Stein, P., & Berliner, D. (1988). Expert - novice differences in perceiving and processing visual classroom stimuli. *Journal of Teacher Education, 39*, 25-31.

Carter, K., Sabers, D., Cushing, K., Pinnegar, S., & Berliner, D. (1987). Processing and using information about students: A study of expert, novice and postulant teacher. *Teaching and Teacher Education, 3*, 147-157.

Department for Education and Science. (1992). *Circular 9/92.* London: Department for Employment and Education.

Furlong, J., & Smith, R. (Eds.) (1996). *The role of higher education in initial teacher training.* London: Kogan Page.

Grossman, P.L. (1992). *The making of a teacher: Teacher knowledge and teacher education.* New York: Teachers College Press.

Haakenson, P. (1994). *Recent trends in global/ international education.* ERIC Digest, October. (ERIC Document No.: ED373021)

Interstate New Teacher Assessment and Support Consortium. (1992). *Model standards for beginning teacher licensing and development: A resource for state dialogue.* Washington, DC: Council of Chief State School Officers.

Jones, L. R., Reid, D. J., & Bevins, S. C. (1997). Teachers' perceptions of mentoring in a collaborative model of initial teacher training. *Journal of Education for Teaching, 23*(3), 253-261.

Kagan, D.M. (1992). Professional growth among preservice and beginning teachers. *Review of Education Research, 62*(2), 129-169.

Kwo, 0. (1994). Learning to teach: Some theoretical propositions. In I. Carlgren, G. Handal, & S. Vaage (Eds.). *Teachers' minds and actions: Research on teachers' thinking and practice* (pp. 215-231). London: Falmer Press.

Lanier, J.E., & Little, J.W. (1986). Research on teacher education. In M. C. Wittrock (Ed.). *Handbook of research on teaching* (pp. 527-569). New York. Macmillan.

Lawlor, S. (1990). *Teacher mistaught.* London: Centre for Policy Studies.

Lucas, C.J. (1997). *Teacher education in America: Reform agenda for 21st century.* New York: St. Martins Press.

McNally, J. Cope, P., Inglis, B., & Stronach, I. (1994). Current realities in the student teaching experience: A preliminary inquiry. *Teaching and Teacher Education, 10*(2), 219-230.

National Research Council. (1996). *National science education standards.* Washington, DC.: National Academy Press.

National Science Teachers Association and the Association for the Education of Teachers in Science. (1998). *Certification and Accreditation in Science Education (CASE) Project.* [On-line]. Available url: http://www.aets.unr.edu/AETS/draftstand.html

O'Hear, A. (1988). *Who teaches the teachers?* London: Social Affairs Unit.

Sanders, L.R., Borko, H., & Lockard, J.D. (1993). Secondary science teachers' knowledge base when teaching science courses in and out of their area of certification. *Journal of Research in Science Teaching, 30*(7), 723-736.

Schön, D. (1983). *The reflective practitioner: How the professionals think in action.* New York. Basic Books.

Schön, D. (1987). *Educating the reflective practitioner: Towards a new design for teaching and learning in the professions.* San Francisco. Jossey-Bass.

Chapter 8

Thinking Like a Teacher

Learning to Teach in a Study Abroad Program

Sandra K. Abell[1] and Amy M. Jacks[2]
[1]Purdue University, U.S.A: [2]Chesterfield County School Corporation, U.S.A..

Abstract: This chapter tells the story of one person's journey toward becoming a teacher. Amy, an undergraduate major in elementary education, participated in a Study Abroad internship in Honduras during the summer following her first year of college. Immersion in the language and culture of Honduras, as well as in the activities of teaching, was a catalyst for Amy's development as a teacher. She engaged in thinking like a teacher as she wondered about student science learning, struggled with discipline issues, and questioned school philosophy and practices. We believe that the Study Abroad internship was a way to make the familiar strange, thus engaging Amy in a level of reflection atypical for novice teachers. We believe her story has implications for increasing the diversity of teaching experiences in science teacher preparation programs.

Learning to teach is a lifelong process. It begins in an apprenticeship of observation (Lortie, 1975) as students watch their teachers teach. However, once students enter a formal teacher education program, a shift occurs. They stop seeing the world solely through the eyes of the student, and begin thinking like a teacher. This chapter tells the story of one such student of teaching, Amy, and the beginnings of her professional development.

INTRODUCTION

I (Abell) first met Amy when she was still in high school, scouting out universities to attend. She planned to major in elementary education and was a science enthusiast (Abell & Roth, 1994)—that led her to my office door. Our next meeting was at freshman orientation; Amy had chosen Purdue as her university. I spoke to the incoming freshman about a project in Honduras, where my colleagues and I were working with Honduran parents and teachers to establish a bilingual elementary school. I mentioned

S.K. Abell (ed.), Science Teacher Education, 141–153.

the possibility of a Study Abroad Program the following summer. After the orientation session, Amy approached me. "I just wanted to let you know that if the Study Abroad in Honduras does happen, I want to be part of it," she relayed.

The Study Abroad did indeed take place during the summer after Amy's freshman year, and we were both part of the experience (Abell as the course instructor, and Jacks as one of four interns). In this chapter we will begin by describing the setting of the story, the Alison Bixby Stone Bilingual School (ABSBS). Next we will explain the nature of the Study Abroad Program. Finally we will examine Amy's experience as a beginning teacher and how she developed her thinking like a teacher in the Study Abroad Program.

THE SETTING OF OUR STORY

Escuela Agricola Panamericana (EAP), otherwise known as Zamorano, is an agricultural college located 25 miles east of Tegucigalpa, the capital of Honduras. The school's 15,000 acres in the Zamorano valley are a fertile spot for growing grains and vegetables and raising livestock. The agricultural college prepares students in a 4-year program that emphasizes "learning by doing," where students spend half of their time in the classroom and the other half in the fields, forests, and local communities. Over 800 students from throughout Central and South America come to Zamorano to participate in what is recognized as one of the finest agriculture education institutions in the Americas.

A pre-kindergarten through 6th grade Spanish-language school, Alison Bixby Stone School, has operated on the Zamorano campus since the 1940's. Over the years, the quality of the school curriculum and infrastructure declined. In 1993, a group of parents, seeking to recreate the school with aspirations toward high norms of quality teaching, organized themselves. The parent association brought forth a proposal to the Zamorano board of directors to create a bilingual school, preschool through grade 6, based on informed educational theory, a state-of-the-art curriculum, and a new organizational structure that would, over time, replace the old school.

Since EAP had collaborated for a number of years with Purdue University faculty members in the School of Agriculture, the Zamorano administration suggested that the parents look for potential collaborators at Purdue. In September of 1994, a team of Purdue School of Education faculty (including the first author), Zamorano parents, and potential teachers conducted a planning workshop to develop the guiding philosophy and framework for ABSBS. This guiding framework was set forth in a document written collaboratively by the Zamorano and Purdue workshop participants (Aguilar et al., 1994). Thus, the school was founded on principles of social equity, cultural pluralism, and experiential learning. Teachers would use strategies such as collaborative learning, multiage groupings, problem solving, and whole language instruction to promote bilingualism/biliteracy in Spanish and English. ABSBS was the first (and to date, only) rural bilingual school in Honduras. It is a private school, as are all bilingual schools in Honduras. Unlike the others, however, ABSBS makes a strong effort to include varied socioeconomic levels among its students, which it accomplishes through a sliding tuition scale and scholarship system. Approximately one-third of ABSBS students receive some form of tuition assistance.

After two years of curriculum development and teacher education at ABSBS, we enlarged the scope of the project by adding a Study Abroad component. Our vision was that Purdue education students, at various points in their teacher preparation programs, would work with ABSBS teachers and their students, and all participants would benefit.

THE STUDY ABROAD PROGRAM

I (Abell) designed the course, "Internship: Schooling, Culture, and Society," to help future teachers "gain experience working in settings that are different from the ones in which they were educated or where they had taught" (Course Syllabus). In particular, the course objectives included:

- Increasing student awareness of a new educational context;
- Improving student ability to synthesize education theory and practice in this context;
- Increasing student skills of participation in an educative community;
- Having students design and teach a series of age and community appropriate science lessons;
- And having students refine personal theories of teaching and learning. (Course Syllabus)

The course, first offered in May of 1996, was divided into 3 parts: 1) for three days we met on the Purdue University campus to become oriented to Honduras and to ABSBS; 2) for three weeks students participated in the internship at ABSBS; and 3) for two days we met back on the Purdue campus to reflect on the experience and our re-entry into the country.

While on campus, students read the first chapters of a field experience textbook geared toward exploring diversity in schools (Powell, Zehm, and Garcia, 1996). These chapters helped orient students to their own beliefs about cultural diversity. Students carried out a number of autobiographical exercises to help them think about their own cultural backgrounds and assumptions. For example, students were asked to describe home and school experiences with persons from different ethnic backgrounds. They wrote an essay called, "What is an American?" Students also gained some background about Honduras and EAP. Finally we explored the ABSBS philosophy; students formulated questions that they wanted to informally investigate during the internship.

In Honduras, students were partnered with an ABSBS teacher and his/her class. For the first week, the interns observed and assisted in their classrooms. Next they collaborated with the classroom teacher to plan a week-long unit of science instruction. During the second week, students team taught with the classroom teacher. By the third week, they taught solo. At the end of the internship, they visited other classrooms at ABSBS and visited other Honduran bilingual schools. The Study Abroad Program also included excursions to local villages, cloud forests, and Mayan ruins.

Students kept journals of their classroom experiences through which I maintained an ongoing dialogue with them. Concurrently they read papers about literacy, mathematics, and science instruction. We met in the evenings as a group to discuss the readings and their classroom experiences. I also met individually with students to discuss their lesson plans and their individual informal investigations of teaching and learning at ABSBS. During the internship, a teacher educator from Purdue University visited ABSBS and conducted a seminar in primary level mathematics education for the teachers and interns.

Upon returning to the US, we took some time as a group to reflect on our experience at ABSBS and in Honduras. Students rewrote their "What in an American?" essays and compared them to the earlier versions. They discussed what they had learned about themselves as Americans, as teachers, and as travelers; what they had learned about teaching and learning; and what they had learned about Honduras and the US. They created an evaluation form for the ABSBS teachers to use to respond to the internship. Finally they evaluated the course itself and made recommendations for the next course offering.

AMY'S STORY

Before the Study Abroad Program

Amy grew up in a university town, the daughter of a university professor/administrator father and a mother who directed children's ministries at their church. She had attended neighborhood schools with other university children, a significant proportion of whom were international students. However, according to Amy, "Most of these kids stayed with people from their own ethnic background" (Course Exercise, 5/13). All of Amy's teachers had been white, and her extracurricular and church experiences were, in Amy's words, "mostly with people of my own cultural group" (Course Exercise). Over the years, Amy's family had hosted numerous international visitors, and so she felt some degree of comfort with people of other cultures. However, her only travel experiences outside of the US prior to the Study Abroad were family vacations to Canada and Mexico.

By the time of the Study Abroad Program, Amy had completed one year of college. In her freshman year, she had taken mainly liberal arts courses, with a strong emphasis on mathematics and science, as well as Spanish. She had yet to take any education courses. She was involved in extracurricular activities through her church and through the campus musical organization. Although she met many new friends in these activities, she tended to stay within her comfort zone and did not enter into activities with a diverse student body. Thus Amy came into the internship with limited background in diverse situations and limited formal preparation in teacher education.

Orientation

During the orientation, Amy expressed some discomfort when it came to interacting with persons from different socioeconomic levels: "I feel that I am not as comfortable as I could be to teach in a multicultural school with different SES, because that's not what I am used to" (Course Exercise, 5/13). However, she was excited to be part of the internship because, "Understanding other cultures will make you more comfortable when interacting with students and make it easier to teach" (Course Exercise, 5/13). In her "What is an American?" essay, she stated that, "An American should be one who is accepting and understanding of all nationalities, because the USA is all nationalities." Thus, Amy undertook the Study Abroad as a way to stretch her experience and grow as a beginning teacher.

After reading the ABSBS philosophy statement, Amy had a lot of questions that she wanted to explore during her internship. First of all, she was curious about the science education in the school: "Are students having opportunities for hands-on, group work, and individual work? Can they apply the science?" (Journal, 5/15). Secondly, she wanted to explore the bilingual aspect of the school: "Are they integrating the two languages too much so that students are getting confused? The sentence structures [of English and Spanish] are different, so are they teaching the differences or just how to translate?" (Journal, 5/15). Lastly she wondered about the social equity stance of the school: "Is the school doing a good job of getting students from different socioeconomic backgrounds? If there is a variety of students, are they all given equal opportunities and treated equally as individuals?" (Journal, 5/15). Amy's questions displayed an attitude of inquiry and a maturity beyond her experience in teacher education. Against this backdrop, Amy entered the internship in Honduras.

The Internship

Immersion in the internship was a catalyst for thinking about teaching. From her first day in Honduras, Amy was asked to think like a teacher:

> *I've only been in Honduras for half a day, but already I feel as if I've learned so much... The students come from so many backgrounds and so their learning is also varied. (Journal, 5/17)*

Amy observed the diversity of students in the pre-Kindergarten class with which she was partnered, and realized she would face some challenges in working with them. These challenges would be derived not only from the diversity of the students, but also from Amy's lack of knowledge and experience.

Amy's beginner status also caused her to feel some anxiety about the internship, a feeling quite common to anyone immersed in a new experience.

> *I'm confident next week will go well, but I'm also a little nervous because I feel I need to incorporate all of the things and new ideas I've been learning about and do them well! I know that I won't be able to do it all, but I need to get over a little bit of the anxiety I'm feeling. (Journal, 5/22)*

Amy entered the internship wanting to perform well, to meet her own expectations and the expectations she perceived from her cooperating teacher and university instructor. She set high goals for herself as a beginning teacher.

Thinking about Learning

Despite Amy's novice status, she showed evidence of thinking deeply about student learning early on in the internship.

> *I also realize the importance of asking a student how they got an answer to understand their thinking. (Journal, 5/21)*

This concern about student learning is rare for novices, who are typically more concerned with personal and management issues (Hall, 1979; Hall and Loucks, 1978). Concern for student learning was not an isolated incident in Amy's journal. For example, she was not satisfied with merely "doing" a science activity with the children:

> *One thing I know I'm learning is using an activity and taking it one step farther. Before this trip, if I would've done the activity we did today with floating paper boats on water, I would have stopped right after the activity was over. Today I was disappointed there wasn't more time because it would have been a great shared experience activity to tie in with literacy. (Journal, 5/22)*

She wanted to be sure she got all the mileage possible from an activity, as a way to build thinking and language skills. Furthermore, she was not easily persuaded by an activity that appeared "fun" to the students:

> *I'm beginning to ask the question, "What did they learn?" from a certain activity. Yesterday [making cardboard cars] they didn't learn a lot... Even though the kids had a lot of fun, it was a lot of hard work for the teachers and the students didn't get to do or learn very much. (Journal, 5/24)*

Amy also struggled with her role as a teacher in terms of student learning. When could she tell a student he/she was wrong? When should she ask another question instead? How should a teacher deal with wrong answers?

> *If their thinking is wrong (not different, because I know there can be different ways to do a problem, but a wrong thinking for the problem), then I don't understand why we can't say their answer is wrong. I don't want to be hard and put a red X on everything, but aren't there some standards in what an answer is? (Journal, 5/21)*

The intensity of the Study Abroad experience required Amy to develop her thinking like a teacher along a faster timeline than is typical of a beginning teacher. However, she had the advantage of daily class meetings, theoretical discussions, and daily oral and written feedback from the course instructor. Such intensity is not common in campus-based courses.

Dilemmas of Discipline

The Study Abroad internship also created some disequilibrium for Amy about discipline issues. Discipline is a concern for most beginning teachers. However, when the disciplinary issues are coupled with diversity issues, new dilemmas are born. Amy spoke several times in her journal about the "wild kids" and the degree of misbehavior she encountered in the preschool class. She had to sort out her own standards for behavior and compare them with the classroom teacher's standards, much like any beginning teacher. However, she had the extra layer of culture to get through in order to understand the situation.

> *Another aspect I focused on during my first day of teaching was discipline. When I took Irene out of the circle for disrupting (biting), the class was in shock. I think the classroom teacher has a greater tolerance for the misbehaving and I don't know if that is a cultural feeling or not. We did talk about how the higher socio-economic children have maids who look after the children. The maids don't have the power to discipline, so children are wild. However, I still don't know if this [misbehavior in the classroom] is due to culture. (Journal 5/28)*

In this case, Amy tried to understand her own action of taking a child out of the circle. Was her action appropriate teacher behavior in terms of age, culture, and classroom norms? At other times, Amy was confident about her disciplinary action. That same day she used a different strategy for dealing with misbehavior and felt comfortable with her decision.

> *Another aspect of my day I noticed was the great amount of tattle-taling. Although the teachers have a great tolerance for the rowdiness in a classroom, they are quick to intervene and solve disagreements. Irene was a trouble-maker in class so when children kept running up to me to tell me what Irene did to them, I just asked them if they would ask Irene to stop, please. At first they didn't know what to do, but then 3 different people at 3 different times asked Irene to stop. It was such a good feeling to know I had influenced children to become more independent. (Journal, 5/28)*

Later in her journal, Amy wrote that "the same form of punishment doesn't work for all kinds of kids," (Journal, 5/31) but that if teachers are both "flexible" and "consistent" with discipline, they will be more successful. One of the factors in helping Amy think like a teacher about discipline was the support of her fellow interns.

> *Discipline is a difficult task to deal with, but I've learned a lot from talking with our Purdue group. I can see how beneficial positive reinforcement is and teaching children to work things out themselves instead of tattle-taling. (Journal, 5/31)*

After the Internship

Upon re-entry into the U.S., Amy and her peers examined their Honduran experience and thought again about what it means to be an American. Amy wrote:

> *I now have a better understanding than an American can be a North American, Central American, or South American. People from the States take for granted the country they come from, and others look at the States as privileged land. (Course Exercise, 6/7)*

Her answer demonstrates taking on another perspective, that is, how a Honduran might view the term American. Taking on new perspectives was an unwritten goal of the Study Abroad Program. In her course evaluation, Amy discussed the many roles she had been asked to play during the internship.

> *I was very pleased with all of the experiences at [at ABSBS]. At times, it was awkward and stressful because we weren't sure about our role. It seems like at times we were Purdue students, then at other times we were asked to observe and assess the school happenings, then we were also "teaching" the teachers... I believe that I really learned a lot about myself as an educator through the experience... I learned to roll with the punches and to be flexible according to the teachers' needs/wants. (Course Evaluation, 6/7)*

Amy had continuing opportunities to think about how the Study Abroad influenced her development as a teacher. For example, in the fall of the year, Amy was invited to at the Dean's Advisory Council Meeting in the School of Education, where she again reflected formally about her experience. At this meeting, Amy stated that her internship in Honduras had helped her learn many things. Her speaking notes delineated several big ideas that formed the core of her learning about teaching.

- [The value of] learning by doing.
- [How to] use local resources for science learning.
- Collaborative learning: Gain respect for and the ability to learn from the ideas of others; Check for their own understanding; Grow in self-confidence
- The Honduras experience brought some of my campus coursework into meaning.
- This experience has given me time in a classroom and has encouraged my life-long learning process. (Notes, 10/17)

Thus, as Amy returned to campus and the regular teacher preparation program, she began to connect the Honduras internship with campus-based coursework, and to find meaning in courses that other students, who had not yet begun to think like a teacher, failed to discover. Amy also returned to Honduras the following summer for a 2-week internship; there she had a chance to test some of the new ideas she had built throughout the school year, comparing them with previous ABSBS experiences.

Amy approached her Honduran experience with an attitude of inquiry, one that was reinforced throughout the internship. Before going to ABSBS, she had questions about how the school functioned, how teachers taught, and how learners learned. The Study Abroad raised new questions for Amy about herself as a teacher. Recently when remembering her time in Honduras, Amy claimed that this questioning attitude was one of the most important components of thinking like a teacher influenced by the Study Abroad.

> *Honduras gave me the confidence that I could do anything! I could work with other educators and stand my own; I could work with children and actually educate them. The experience in Honduras was the most challenging adventure I had ever taken on, and I did it. Because of that experience, I had the confidence to question my professors, peers, and students. I don't feel as if I went through college to learn everything about teaching. Instead, college allowed me to take the material I learned from my professors and peers and make sense of it through questions, analysis, and discussion with others. Because I was "immersed" into teaching in Honduras I had to ask questions to survive. Rather than be spoon-fed during my methods courses, I was equipped, because of my Honduras experience, to ask questions and learn from my questions. This helped me in student teaching too. Students learn more if they get a chance to be "immersed" in their own learning. Let them ask questions without giving them all the answers. Productive questions [that we learned in science methods (Elstgeest, 1985)] totally make sense to me because I see the importance of asking and utilizing questions. Questions take you to another level of learning. (Reflection, 1999)*

Amy claimed that her Honduras experiences have influenced all of her teaching situations. The most recent examples come from student teaching. During student teaching, Amy used Spanish with 2nd graders as a way of developing their literacy skills. For her 5th grade rotation, she used her episode (and her photos) of hiking in a Honduran cloud forest, to support teaching about rainforests. The Study Abroad Program thus helped Amy develop knowledge, dispositions, and skills (cf., Interstate New Teacher Assessment and Support Consortium, 1992) for becoming a teacher.

CONCLUSION: MAKING THE FAMILIAR STRANGE

An immersion experience is a way to make the familiar strange. The commonplaces of our lives are challenged and we can think about issues in new ways. This premise forms the foundation of many study abroad programs (Moyars, 1994). For students of teaching, making the familiar strange is not always easy. They have sat through twelve or more years of their apprenticeship of observation (Lortie, 1975) before enrolling in a formal teacher preparation program. Schools seem all too familiar, and teaching and learning appear non-problematic. Making the familiar strange can challenge students of teaching to think about schooling in new ways. The Study Abroad Program is one vehicle for making the classroom a less than familiar setting.

For Amy, the Study Abroad internship was an immersion experience on several levels. First, she became immersed in a world that was new to her. The setting was different, the food was different, and the language was different.

> *I found it difficult to understand when a child couldn't understand my English, or if they couldn't understand patterns [the lesson's topic]. (Journal, 5/28)*

This confusion meant that Amy could not take her communication for granted. Interactions with children took on a significance that they might not have back in the States. Wondering about the role of culture in terms of discipline, or the role of language in teaching and learning, forced Amy to think in new ways about everyday occurrences.

Amy was also immersed in the culture of schooling. She had been taken out of her student role and placed in a setting where children looked to her as teacher all day every day for three weeks. With little formal preparation in teaching, she was forced to use the various resources at her disposal—the classroom teacher, the course instructor, other interns, course readings—to make sense of the experience. Perhaps this immersion, coupled with on-site support, allowed her to think more deeply about student learning than is typical for novices. Thus the internship could provide a firm foundation for future education courses.

Amy left her comfort zone to participate in the internship. She reflected on her experience in this way:

> *My immersion experience, both culturally and professionally, forced me to be uncomfortable at times. Culturally, I was "forced" to use Spanish, slow down my daily pace, and be flexible. The professional aspect of my immersion program allowed me to develop as a professional educator. I*

> *was "forced" to interact in a professional manner with parents and teachers. I learned to develop classroom management skills. I learned what it means to ask thought provoking questions, take a lesson one step further, and to implement my learning about educating children into my daily practices. (Reflection, 4/99)*

We believe that learning to teach is a developmental process that takes place across one's teaching career. The role of an immersion experience is to create disequilibrium, to make the familiar strange. It can be transformative—helping a novice begin to think like a teacher—but it is not instantaneous. Amy returned to her teacher preparation program where she spent three more years taking methods courses and having field experiences in local schools. Looking back on her internship journal nearly three years later, she wrote:

> *My journals have an innocent and a naïve sound. I believe this is due to my lack of curriculum knowledge and educational training. [During the internship] I was allowed to develop and grow. I think this says something important about getting students out into the classrooms early on. (Reflection, 4/99)*

We believe that Amy's journals also say something important about the benefit of diverse experiences in developing teacher thinking. Although it is unrealistic to think that every future teacher will travel to another country to experience an internship, there are ways to make the familiar strange closer to home. Students could have experiences as teachers in school settings different from the ones in which they grew up. Students could also have experiences as teachers at grade levels different from the ones in which they intend to teach. A diversity of teaching experiences could help to highlight issues of teaching and learning that might not be uncovered unless the familiar is made strange.

REFERENCES

Abell, S. K., & Roth, M. (1994). Constructing science teaching in the elementary school: The socialization of a science enthusiast student teacher. *Journal of Research in Science Teaching*, *31*, 77-90.

Elstgeest, J. (1985). The right question at the right time. In W. Harlen (Ed.), *Primary science: Taking the plunge* (pp. 36-46). Portsmouth, NH: Heinemann.

Hall, G. E. (1979). The concerns-based approach to facilitating change. *Educational Horizons*, *57*, 202-208.

Hall, G. E., & Loucks, S. F. (1978). Teacher concern as a basic for facilitating and personalizing staff development. *Teachers College Record*, *81*, 36-53.

Interstate New Teacher Assessment and Support Consortium. (1992). *Model standards for beginning teacher licensing and development: A resource for state dialogue.* Washington, DC: Council of Chief State School Officers.

Lortie, D. C. (1975). *Schoolteacher: A sociological study.* Chicago: University of Chicago Press.

Moyars, K. (1994, Summer). A global experience. *Purdue Perspective.* [On-line]. Available: http://www.ippu.purdue.edu/sa/general/Publicity/perspect1.htm

Powell, R. R., Zehm, S., & Garcia, J. (1996). *Field experience: Strategies for exploring diversity in schools.* Englewood Cliffs, NJ: Merrill.

SECTION IV

CROSS-CULTURAL PERSPECTIVES ON SCIENCE TEACHER EDUCATION

Chapter 9

A Meeting of Two Cultures

The Experience of Facilitating a Teacher Enhancement Project for Egyptian High School Science Teachers

Janice Koch[1] and Angela Calabrese Barton[2]
[1]Hofstra University, U.S.A.: [2]University of Texas, U.S.A.

Abstract: In this chapter, we explore the experiences of two science teacher educators participating in a Binational Fulbright Commission Teacher Training Initiative with eight public school secondary science teachers from Cairo, Egypt and neighboring communities who came to the United States to study science teacher education. In the first part of the chapter, we examine the preparatory nature of the program. We then identify and analyze the key players in this science education initiative in terms of their assumptions about teacher education, science, and cultural difference. The last section of this chapter describes the reflexive nature of the relationships between teacher educators and the participants in the context of a binational program. The chapter is guided by the following questions: (1) What assumptions were made about the visitors before they arrived? (2) How did perceived cultural differences inform the structure of the program? (3) How did actual cultural differences transform the science methods course? (4) How were the teacher educators transformed by this experience?

This chapter explores the experiences of two science teacher educators participating in a Binational Fulbright Commission Teacher Training Initiative. Eight public school secondary science teachers from Cairo, Egypt and neighboring communities were part of a larger group of secondary teachers who visited a private university in a suburb of New York City. The 3-month project included providing the visiting science teachers with a science methods course and a related field placement experience. The language of the grant stated that the visiting teachers would gain skills in reflective pedagogy, activity-based learning, constructivist approaches to science education, and cooperative learning.

This chapter is informed by the conversations and observations of the teacher educators who designed and taught the science methods course for

S.K. Abell (ed.), Science Teacher Education, 157–170.

the visiting teachers. These data impact our understanding of cultural difference and the implications of cultural difference for communication about science education. The authors of this chapter were not the Principal Investigators for the grant and were not responsible for creating the structure in which they were required to work with the visiting teachers from Egypt. These disclaimers are important, since the structure and affective climate into which these Egyptian teachers entered had an underlying subtext that became a source of tension for the teacher educators. These tensions between the socio-cultural context in which this project was situated and the socio-cultural context of the visiting teachers informed much of what was learned in the science methods course. Some of the issues we address include:

- What assumptions were made about the visitors before they arrived?
- How did perceived cultural differences inform the structure of the program?
- How did actual cultural differences transform the science methods course?
- How were the teacher educators transformed by this experience?

In this chapter we will explore these questions. In the first part of the chapter we examine the preparatory nature of the program. We then identify and analyze the key players in this science education initiative in terms of their assumptions about teacher education, science, and cultural difference. The last section of this chapter describes the reflexive nature of the relationships between teacher educators and the participants in the context of a binational program. This section is informed by the following sources of data: email communications between the two science teacher educators, personal narrative and essays written by the participants, participant observation, and Binational Fulbright commission material. The autobiographical data and our participant observations provide a thick matrix from which the actual experiences are described. These data impact our understanding of cultural difference and the implications of cultural difference for communication about science education.

THE BINATIONAL FULBRIGHT PROGRAM

The Binational Fulbright Commission Teacher Training Initiative: Inservice Teacher Enhancement for Egyptian Mathematics and Science Teachers (Funding Proposal, 1996) was a 3-month immersion experience for 30 public middle and high school science teachers from the greater Cairo, Egypt area. Of the group, there were eight public school science teachers from Cairo and neighboring communities. The grant was designed, directed, and administered by a team of administrators from a private university in a suburb of New York City in collaboration with the Fulbright Commission's Egyptian based project director. The program occurred on the campus of the university and in two major school districts nearby; these districts were pre-selected by the project directors based on their proximity to the university and their willingness to have the Egyptian teachers participate in their schools. The science teacher educators were selected from the full time faculty at the university after the grant was funded.

Common to Binational Fulbright Commission Teacher Training Initiatives, as stated in the commission's literature (Binational Fulbright Commission, 1995, pp.3-4), the goals of the program were:

- To develop and to enhance the participants' use of spoken and written English, particularly in interactive classroom situations.
- To develop and expand the participants' knowledge base and repertoire of teaching techniques and methodologies, especially as related to the skill areas of reading, writing, listening, and speaking.
- To develop and to enhance the participants' classroom management skills.
- To introduce the participants to computer literacy/word processing skills and to computer assisted learning.
- To familiarize the participants with a culture in which English is the native language through a total immersion experience, so that participants have direct exposure to issues such as cultural diversity, conflict resolution, and participatory decision making processes.

In addition, the objectives of the university's program were to provide experiences that developed the teachers' ability to:

- help students understand and apply concepts rather than memorize rules and procedures;
- use concrete materials and real world applications;
- use experiential learning and cooperative learning activities;
- use community resources;
- integrate knowledge from multiple discipline knowledge bases;
- improve or enhance the teachers' spoken and written English, computer literacy, classroom management skills, and knowledge of American culture.

As stated above, the language of the grant required the science teacher educators to prepare the visiting teachers in specific methodologies. Hence, despite the fact that cohort group members were practicing teachers in Egypt, there were expectations embedded in the written proposal that their education reflect beginning teacher preparation.

In this framework there were three key groups of players: the Co-Principal Investigators are the project directors; the authors of this chapter are the science teacher educators; and the participants are the visiting Egyptian teachers. This triad experienced this teacher education program in distinctly different ways. Given this framework, the project directors designed a program with four major strands: (1) intensive coursework in science teaching; (2) remedial work in English; (3) field based experiences (participant observation and teaching); and (4) workshops in technology and alternative assessment. This meant that for the first month of the 3-month experience, the participating teachers spent Monday through Thursday mornings in an intensive science methods course, the early afternoon in a technology education course, and the evening in remedial English courses. The next six weeks of the course involved an intensive field experience. For this field experience, each teacher was placed with a cooperating inservice teacher of the same subject area(s) and grade level(s). The teachers spent six hours per day in the schools as participants and observers, Monday through Friday. During this 6-week period, the teachers continued to spend their evenings in remedial English courses. The last two weeks of the program involved a group video project where the teachers documented "what they learned" during their experience. It is important to note that the project directors believed it was in the best interest of the Egyptian teachers to have full, programmed days with little free time. This schedule was arrived at by the project directors in collaboration with the Egyptian project director. Neither the participants nor the science teacher educators had a voice in the design of the program (the strands, sequence, or daily schedules). However, the science teacher educators had control over the design of the science education portion of the program. What follows is a description of the science education experience (methods and field-based) that was embedded in the overall program context.

FRAMING THE SCIENCE METHODS COURSE

The hierarchical division of labor described above proved to be dissonant with the approach and belief systems of the two science teacher educators. It was also a source of tension for the Egyptian teachers who had entered this program with extensive teaching experience and (as later learned) an articulated set of teaching ideals to be explored and refined, but as outsiders to the program and to the American university in which the program was housed.

For us, the two science teacher educators, the tension resided in how we perceived science and teaching and how science and teaching were perceived in grant. As feminist science educators, we were committed to reflect upon our own and our students' experiences with science and teaching as a basis for our work with them. We recognized that our work with the project was deeply connected to how much we knew about our students, their experiences, and their homeland. We knew that their teaching beliefs and ideals might be different from ours (in terms of day-to-day practices as well as long term goals) because we have been educated in different systems. Finally, we believed that although we had expertise in science education, we could not, at the outset of our work with teachers, know, definitively, what they would "need" in their American educational experience. These questions and circumstances focused our discussions around what we might do in our teaching with the Egyptian teachers, even if that was not what was intended in the proposal.

Our story begins with the planning process, as both science teacher educators sat in a small office one early spring morning constructing a syllabus for the experienced science teachers from Egypt with whom we would soon be working. What can we discover about each other? What are our beliefs about the nature of science and science teaching? What are the implications of integrating a feminist critique of natural science and multiple ways of conceptualizing science?

Informing much of what we planned for the teachers was the belief that science education had somehow evaded multiculturalist critique by appealing to a universalist epistemology: that the culture, gender, race, ethnicity, or sexual orientation of the knower is irrelevant to scientific knowledge (Stanley & Brickhouse, 1995). Believing personally that this was not the case, we invited our students to write their science autobiographies (Koch, 1990). Later in this chapter, excerpts from these autobiographies introduce the science teachers from Egypt to the reader. Furthermore, the autobiographical data were used to modify the design of the methods course by weaving the teachers' stories into our course content.

THE COURSE

The science methods course was constructed around two themes.

Ways of knowing science (as teachers, students, and scientists):

- Reflecting on the personal construction of the nature of science.
- Building a personal vision and philosophy for teaching science.
- Examining who does science and why, and who does not do science and why not.
- Exploring the role that science class plays in developing students' ideas and attitudes about science.

Developing an inclusive and constructivist based science teaching practice:

- What is the image of science we want to construct in classrooms?

- What is the connection between school science and society?
- How do we develop a knowledge of traditional American secondary school science curriculum?
- What do science teaching models have to say about the nature of science, the nature of teaching and learning, knowledge construction, and students as scientists?
- How do we develop strategies for planning units and lessons in science that foster classroom discourse that values students lived experiences and that engages students in really making sense of the world around them?

To engage meaningfully in these themes, the course participants wrote science autobiographies, interviewed students, reviewed national and state curriculum frameworks and local curriculum guides, and developed their own presumably constructivist based lessons and activities.

BUILDING COMMUNITY ACROSS DISCOURSES

Our first meeting with the Fulbright teachers was at a large group luncheon. Only two of the science teachers were at our table for lunch, but we had a wonderful time chatting with each of them. They were women from Egypt, on the road, far from home, and science teachers. Their smiles appeared to us as warm and genuine and we were anxious to meet the entire group of eight science teachers. We believed that, if these two women were exemplars of their group, that we would have no problem at all communicating. This surprised us because we had been informed that the women teachers had low English proficiency scores. This experience forced us to begin to question how the Egyptian teachers--and their educational needs--had been positioned in the project. Knowledge of the teachers' English proficiencies encouraged our culture to marginalize the teachers and to set a low level of expectations about their performance that pervaded the supervisory atmosphere. Clearly the university teacher educators were supposed to be "giving them" the conventional wisdom about constructivist science, which it was assumed they sorely needed. At no time was it suggested that our guests could offer the professorate meaningful experiences. In the words of Elizabeth Minnich (1990), "Whiteness has been claimed as justification for dominance [and superiority] over others" (p xvi).

On the first day of class, the eight teachers from Egypt entered the science lab looking dazed and a bit confused; there were five women and three men. They had been teaching science for several years in Egypt, in areas including physics, chemistry, general science, and biology. They were more than a little upset that their living conditions were less satisfactory than they had hoped for--they disliked sharing bathroom facilities, the dorms were old and musty, and they felt cramped and uncomfortable with these conditions. We felt embarrassed that the university could not have found more suitable lodging. As time went on, our colleagues from Egypt sorely missed food from their home and we arranged occasions to bring in Middle Eastern food.

Using personal experience to examine science was central to our efforts to create a science learning community across discourses. This process helped us to uncover the power, limits, and partiality that inscribe our sense of identity in science and education, our relationships with the people acting within our institutions, and our own vision for change. More importantly, using the personal forced us to embrace our experiences with a critical eye so that we could find ways to connect with our Egyptian colleagues.

At the first meeting with the science teachers, we asked them to write their science autobiographies. In what follows, we share excerpts from the stories of three of the nine teachers and our reflections on them. We believe that the underlying themes emergent from these stories is characteristic of each of the stories presented by our Egyptian colleagues. More importantly, we selected these particular stories because we felt they helped us as teachers and researchers to understand, in more poignant terms, the concerns and issues experienced and expressed by our colleagues. Therefore, although we believe the themes presented in these stories are common to all of the teachers, we also recognize that the intensity in which the themes are presented in these stories is not necessarily generalizable.

Through the autobiographical process, we established connections and shared stories. The result of this process was that we learned things about each other and about science that were not aspects of what was intended to be taught and learned as part of the teacher preparation grant. In addition, we were struck by how facile our Egyptian students were in writing English; we reflected upon our own inability to express ourselves in their native tongue.

It is important to note that the use of the science autobiography in written narrative form was the first exercise we implemented to come to know the Egyptian teachers and understand more about their personal and professional values. They wrote their stories as a written homework assignment and handed them in at our second meeting. We used their stories to help frame the direction of the course. We wanted to learn more about our students as individuals and we wanted them to reflect on their own histories as a way of understanding their own positions as science educators and how they came to frame their teaching values.

The first autobiographical excerpt was written by Mahmoud Abdel Fattah, a high school physics teacher, who appeared to be a serious man in his late twenties with a strong sense of urgency about his participation. One of Mahmoud's major reasons for joining this group of inservice teachers was to learn as many American strategies about teaching physics as possible. He talked often about how important it was for him to take these skills back to his students in Egypt. He was influenced by his own teachers and the following story illustrates a pivotal moment in his professional pursuits.

> Mahmoud Abdel Fattah:
>
> *My considering with science started when I was in preparatory stage. Once a day there was a science test, and after my science teacher finished from the correction of test, my degree [grade] was always 10.5 out of 12.. . I started to cry because I felt shy; I like my teacher very much and I was afraid that my picture [image] change in the view of my teacher. When she noticed me, she came to me and try to get me quiet and showed me where my mistake and encourage me to get the full degree in the next test. At this instant I felt this teacher [to be] the greatest thing in my life. From that day my science teacher became a good model for me and I try to be good teacher like her.*
>
> *My love of science continued with me also in secondary school and one day my physics teacher was absent. So some students and I decided to occupy this period by studying physics together. They asked me to explain a certain point on this class board. At this instant the vice principal of school came in the class and noticed me, but he thought that I laugh on my teacher by imitating him, so he hit me on my face. . . I felt angry but after a little time I went to his office and explained to him why we did this thing. He was sorry to me and encourage me from that day to be a good teacher.*
>
> *I believe we must teach science as a part of our life. Science help me feel more capable to deal with everything through understanding. Science is me and I am science.*

Getting to know Mahmoud through his science autobiography helped us when we were later confronted with a problem in the field placement. As part of the project, the Egyptian teachers were placed in high schools in a middle class white neighborhood in western Long Island. Each day they were bused to and from the schools so that they could experience "American education" first hand in their own subject areas. However, Mahmoud, a physics teacher, was placed in an Earth science classroom rather than in a physics classroom. When he repeatedly requested to the project director to be moved into a physics classroom, he was told that "if he did not like his Earth science placement, he could pack his bags and return to Egypt." We were told that Mahmoud was a "troublemaker" when, in fact, we knew him to be sensitive and deeply committed to his profession. What prompted him in his persistence about the proper placement was his own sense of urgency about this field experience. Although we eventually found Mahmoud a physics classroom (against the project director's decision), he and the Earth science teacher found exciting ways to create a short astronomy unit integrating physics and Earth science.

As we read Mahmoud's story, our belief in the power of human narrative re-emerged. Much has been written about the stories lives tell in educational practice and research (Witherell & Noddings, 1991). In all the teachers' stories, the common theme that emerged was what Maxine Greene refers to as "the ways in which stories give shape and expression to what otherwise would be untold about our lives"(in Witherell and Noddings, 1991, p. 3). Reading Mahmoud's science autobiography revealed a personal and sensitive, yet pivotal experience for Mahmoud and crossed cultural boundaries. While the physical "hit on the face" may not be usual in our cultural experience, there are metaphoric "slaps on the face" that potentially shut the doors of scientific study for many.

The second science autobiographical excerpt was written by Mary William Saad Salama, a general science teacher in a preparatory school that addressed grades that are the equivalent of our middle schools. She was a woman in her mid-twenties who dressed in traditional Middle Eastern clothing and had a bright, cheery smile at all times. It appeared that Mary was very happy to be part of this inservice teacher group and was eager to participate. Similar to Mahmoud's autobiographical excerpt, Mary's excerpt raises questions about access to science. Mary's autobiography reminds us how sharing personal experiences provides us with a safe space to approach the goal of questioning the ways in which authority can be empowering or disempowering.

Mary William Saad Salama:

The science and its branches are the most favorite subjects to me. I do my best to simplify it to become easy to understand by all students through all stages. In primary stage, I was loved by my science teacher. I was encouraged by the teacher to do simple experiments at home about states of matter, melting of ice, boiling of water, evaporation by heating and burning of candles. Teachers dealt with us friendly so we loved them and loved science. When I entered the college of science. . . I got an awareness about science, environment, research and the application of science in life. I take my students to the garden and let them cultivate seeds to notice the growth, and notice the importance of green plants as a source of oxygen and food. . .so [they] don't pick the flowers. I take them to the beach to collect snails, algae, and sea animals. . . I enjoy teaching very much and I transfer the importance of science for them because I think the world will be controlled by powerful people who know more about science.

Mary's autobiography shows us how authority can be used in positive ways to empower students. Her science teacher encouraged her to explore the natural world outside of the context of school. She saw science as informing power structures and believed that being empowered through science made her less vulnerable to outside authority. In the story that follows, we saw Mary questioning authority in the public sphere because she was empowered to do so. Several weeks into the project, Mary invited us to her dormitory room because her electric shaver from Egypt was not working. Mary joined us, along with two of the other women teachers from Egypt, as we drove to various electronic stores buying what we needed to get her electric shaver to work, once again. Mary was not persuaded by the cool talking salesman that what she really needed to do was to buy a new electric shaver. By the end of the afternoon, with no help from the outside world, we had her shaver working once again. This experience was a bridge between our worlds, not only linking our common interest in understanding how the world works, but also linking our desire to be "strong women," women not easily persuaded by those hoping to dominate.

Furthermore, these stories also reveal our attempts to thwart the ways in which the Egyptian teachers were inundated with the goals and objectives of the project, because we felt this practice only served to maintain their social and cultural marginalities. Instead, we believe, learning became a counter-hegemonic act, "a fundamental way to resist every strategy of white racist colonization" (hooks, 1994, p. 2). For example, when Mahmoud refused to accept his unacceptable situation, he found ways to use his expertise in physics to engage in American education. In the project's attempts to silence and marginalize this man, as well as to keep him from interacting meaningfully with American students, he resisted, and found ways to connect to the students. In Mary's story, she would not use her cultural marginality to be forced into making an unnecessary purchase. Both cases highlight the self-esteem and competence of the teachers.

The third autobiographical excerpt is from Ibtesam Mohamed, a secondary school biology teacher in her late thirties. Although a quiet woman in social contexts, Ibtesam came to life when she was teaching. She dressed in traditional Middle Eastern clothing and spoke often about her three children who were at home in the Cairo area. She missed her family and contacted them by telephone often. However, she was determined to derive as much professional gain as possible from this trip to the US. Participating in this program was so important that she was willing to leave her family for 12 weeks. Because of her age and her teaching experience, she was looked upon as one of the leaders by the other teachers in the group. As her autobiographical excerpt explains, her teachers engaged her in doing science lessons even before she completed her education. Her teachers played an important role in helping her find her love of science and the natural world.

Ibtesam Mohamed:

I had wonderful teachers of science in preparatory school. They took me to science lab and wanted me to share with them my expectation about the results of experiment[s]. From this stage I liked science and in the secondary stage I joined with the Biology section and my teacher asked us to prepare a lesson for the class. With the passing of time I taught and also learned science. Actually I love the nature very much and science helps man to discover the environment and realize the wisdom of God to create everything.

Ibtesam's autobiography reinforced her innate love of nature and her desire to share that with students. In their field experience, the Egyptian teachers were assigned the role of "observer." They indicated how much they missed teaching their students and they were frustrated about not having a voice in their own schedule. They were rarely asked to contribute to this experience and, hence, they were not valued as knowers. We invited those who wanted to teach a lesson in their specialty area to work with students in an undergraduate section of the elementary science methods course. They were also asked to contribute to the graduate student teaching seminar and share their own science teaching wisdom.

Ibtesam volunteered to teach a lesson on the structure of the flower and we picked up several stems of gladiolas for her. She captivated the students with her appreciation of the flower and her willingness to share the flowers with the class and enable them to connect form and function. The students benefited from the lesson and gained an opportunity to teach the structure of the flower lesson with their own elementary students in their co-requisite field placement.

The students were engaged with Ibtesam in a way that bridged their perceived cultural gap. The undergraduate students spoke at length about how enthusiastic she was and how clearly they understood her lesson. They expressed surprise during this discussion and we were able to reflect upon our preconceived notions of those who appear different from our culturally constructed image of the "teacher." Their tacitly held beliefs about the visitors from Egypt as being lesser or other were deconstructed by Ibtesam's desire to teach them. It was Ibtesam's lesson, permeated with her warmth and her connection to nature, that engaged them and altered their preconceived notions of the science teachers from Egypt. It was a more important learning experience than any they had that semester.

We believe that our awareness of the Egyptian teachers' lives as well as the cultural politics at play, allowed us to know the teachers in ways that enabled us to read their actions in productive rather than regressive ways.

DISCUSSION

As scientists, educators, and teachers, we are faced with the challenge of teaching and understanding the "other." In science education, we are confronted with the "other" through constructs of gender, class, ethnicity, race, and sexual orientation. People who do not meet the traditional and stereotypical image of a scientist in physical or psychological terms are immediately labeled as outsiders, or at best marginal participants in science. Nowhere is this marginalization more obvious than in the way the science education community writes about those who are in need of remedial or compensatory programs. This kind of talk is evidenced in the science education literature when researchers intervene on behalf of underrepresented populations in science by suggesting ways to bring their performance "up" to standard. This is in contrast to approaching the perceived "other" as a contributor to the science education context, not as an individual who needs to be "fixed."

For example, the Egyptian teachers were at once present and invisible in the educational discourse of the project. They were recognized to the extent that fragments of their identity had become an issue for political debate (i.e., their facility with English). They were invisible to the extent that they were never consulted about the construction of their schedule or educational experience. Their views about science teaching and learning were not central to the goals of the project. An interesting paradox in science education is the contradiction between discourses that call for recognition of students' identities while they promote a universal notion of science or science education that trespasses any boundaries of race, ethnicity, class and gender.

We recognize that interrogating difference is a source of growth and revisioning (Luke & Gore, 1992), but we also recognize that rendering difference as the only visible and viable analytic category promotes a hierarchical schematic that perpetuates systems of marginalization. This hierarchical scheme is driven by the desire to twist the unfamiliar, the foreign, and the "incompatible" to the outlandish, the inferior, and the deviant. Our stories illustrate how easy it is for those in power to use the label of otherness as an excuse to dominate. bell hooks (1994) writes that many black students are not taught by teachers who believe that educating "black children rightly ... is a political commitment" (p. 2). Instead, students of color are often exposed to an education that reinforces racist stereotypes.

What was happening at our university was no different than what is happening within the larger science education community. Contemporary science education reform efforts in the United States have placed an emphasis on the importance of scientific literacy and understanding for all students, not just those who may pursue careers in science (American Association for the Advancement of Science, 1989, 1993; Goals 2000, 1994; National Research Council, 1996). The movement towards scientific literacy has been fueled by an understanding of science as a social construction (Harding, 1986, 1987, 1991; Keller, 1985; Longino, 1990). The movement towards science as a social construction is important when considering the challenge of involving all students in science. Not unlike science, science education, has also been grounded in the positivistic assumption of an objective reality (Barton, 1998; Brickhouse, 1994; Stanley & Brickhouse, 1995). Despite the fall of positivism in philosophical circles, many science educators use their authority to engage students in dialogue that emphasizes objectivity and rationality. This pedagogical discourse requires students to make sense of the world in prescribed ways. They learn to impose boundaries, constraints and definitions on themselves, others, and the world. Consequently, students also learn that if the prescribed ways for engaging in science do not "connect to their experiences," then the student is the one who is intellectually or culturally deficient. Knowledge construction and information organization are assumed to be immune from a priori cultural assumptions of those carrying out the work and from the political, governmental, and national interests that guide and fund the work (Weinstein, 1998). From this framework, the construction of scientific knowledge is perceived to be influenced only by that which happens within—inside—the community of scientists. This leaves unchallenged the traditionally valued knowledge, authority and power of science that undergirds relationships between science, teachers and students.

Clearly the expectations of the directors of this grant, of ourselves as teacher educators, and of the Egyptian science teachers needed to be examined, exposed, revisioned, and compromised in order to achieve the kind of mutual communication that leads to true learning. The exploration of this experience has important implications for how we frame our ways of being as teacher educators. We believe that we need to develop critical understandings of how we know what we claim to "know" in science education. We also need develop critical understandings of the assumptions that we bring to knowing. Finally, we need to work with students so that they may develop such critical understandings as well.

REFERENCES

American Association for the Advancement of Science. (1989). *Science for all Americans.* New York: Oxford University Press.

American Association for the Advancement of Science. (1993). *Benchmarks for scientific literacy.* New York: Oxford University Press.

Barton, A.C. (1998). *Feminist science education.* New York: Teachers College Press.

Binational Fulbright Commission. (1995). *Teacher training initiative (TTI). Request for proposal, integrated english language program.* Garden City, Cairo, Egypt, November 23, 1995.

Brickhouse, N. (1994). Bringing in the outsiders: Reshaping the science of the future. *Curriculum Studies, 26* (4), 401-416.

Funding Proposal. (1996). *Teacher training initiative (TTI): In-service teacher enhancement for Egyptian mathematics and science teachers* (A proposal submitted to the Binational Fulbright Commission's Integrated English Language Program.)

Goals 2000: Educate America Act Pub. L. No. 103-227 (33/31/94), Stat. 108 (1994).

Harding, S. (1986). *The science question in feminism.* Ithaca: Cornell University Press.

Harding, S. (1987). *Feminism and methodology.* Bloomington: Indiana University Press.

Harding, S. (1991). *Whose science? Whose knowledge? Thinking from women's lives.* Ithaca: Cornell University Press.

hooks, b. (1994). *Teaching to transgress: Education as a practice of freedom.* London: Routledge.

Keller, E.F. (1985). *Reflections on gender and science.* New Haven: Yale University Press.

Koch, J. (1990). The science autobiography project. *Science and Children, 27*, 43-45.

Longino, H. (1990). *Science as social knowledge: Values and objectivity in scientific inquiry.* Princeton: Princeton University Press.

Luke, C., & Gore, J. (1992). Introduction. In C. Luke & J. Gore (Eds.), *Feminisms and critical pedagogy* (pp.1-14). New York & London: Routledge.

Minnich, E. (1990). *Transforming knowledge.* Philadelphia: Temple University Press.

National Research Council (1996). *National science education standards.* Washington, DC: National Academy Press.

Stanley, W., & Brickhouse, N. (1995). Multiculturalism, universalism and science education. *Science Education, 78*(4), 387-398.

Weinstein, M. (1998). Playing the paramecium: Science education from the stance of the cultural studies of science. *Educational Policy, 12*(5), 484-496.

Witherell, C., & Noddings, N. (1991). *Stories lives tell: Narrative and dialogue in education.* New York: Teachers College Press.

Chapter 10

International Partnerships as a Means of Reforming Science Teacher Education

Bolivia, Chile, Venezuela, and the United States

Jon E. Pedersen[1], Ronald J. Bonnstetter[2], Marilu Rioseco[3], J. Mauro Briceno-Valero[4], Hernan Garcia[5], and James O'Callaghan[4]

[1]University of Oklahoma, U.S.A.: [2]University of Nebraska, U.S.A.: [3]Universidad de Concepcion, Chile: [4]Universidad de los Andes, Venezuela: [5]Cochabamba Cooperative School, Bolivia

Abstract: This chapter examines, in the context of individual countries, the reform efforts that are ongoing in the area of science teacher education. It also examines the simultaneous renewal process of the development of partnerships and the mutual assistance and growth that these partnerships offer. This chapter does not profess to provide a "model" that can be easily implemented in other institutions. Rather, we hope that readers will take away information about how these partnerships have been arranged, the impact of the partnerships on all stakeholders involved, a better understanding of the change process in diverse settings, as well as several case study overviews of reform efforts and their impact on the countries involved.

The diversity of public and private education around the world is immense. Our experiences in international education have taken us from the mountains of Venezuela to the southern coast of Chile to the jungles of Bolivia. We have embarked on odysseys in which we examined urban settings and walked for miles to reach an unrecognizable setting called a classroom. You have to place yourself in these situations to appreciate the sheer magnitude of the contrast of "perceived" teaching environments in which learning is to occur.

There are a significant number of individuals and classrooms out there that need, and indeed are asking for help. For example, imagine walking into your classroom and seeing no light switch, no overhead projector, no Internet connection, and no equipment of any kind. Imagine there is no electricity, no clean running water, or books in the classroom. Now, imagine that the groups of children that you are teaching do not speak the same

S.K. Abell (ed.), Science Teacher Education, 171–192.

language that you do or hold the same cultural values as you. This may seem extreme, but it is a reality for many science teachers in many countries throughout the world. In many instances, in classrooms throughout the northern and southern hemisphere, windows are broken, buildings are infested with termites, and children squeeze into rooms that are overcrowded and ill equipped. Add to this milieu the fact that many of the science teachers entering these classrooms are ill prepared in science content and pedagogy, as well as ill prepared to deal with the extremes present in today's classrooms, and one can see why failure can become commonplace.

ECONOMICS, POLITICS, AND EDUCATION

An examination of economics and regional politics provides many insights as to why these deplorable situations have arisen in South America and throughout the world. For example, the economic growth model, used for the last 50 years in many of these countries, is based on reducing imports by stimulating industrial growth. This has led to an exodus from the countryside to the cities, thus upsetting the urban equilibrium and causing deterioration in the availability and quality of public services at every level. Additionally, for many Latin American countries, the advent of democracy over the last 40 years was built on promises by politicians for a better standard of living. At the same time, education became the most important instrument for social mobility and a means for reaffirmation of the declared principles and practices of these newborn democracies. This resulted in notable increases in student numbers at all levels of education. The democratic regimes provided mass education, but of an inferior quality. For example, the increase in student numbers led to what is known as the "double-shifts," in which one group of students uses a facility in the morning and another in the afternoon. At the same time, the lack of a systematic policy for preparing teachers or providing refresher courses has meant that unqualified personnel have entered the system, making the crisis worse than ever.

Fiscal irresponsibility also has led to serious internal problems within these countries. In the 1980s many countries in South America had been plunged into the worst economic crises in their history, marked by currency devaluation and rampant inflation. This led to predictable social consequences--unemployment, malnutrition, social confusion, and uncertainty--followed by poor school attendance and decreased academic achievement. Far too often, educational institutions have been held responsible for poor academic standards and the social irrelevance of their courses without confronting the underlying problems.

SCIENCE EDUCATION

All across the world, political, social, and economic issues are driving educational systems to push for reform. In North and South America, the current battle cry is no different. Because of such problems, current educational systems have dominated the horizon for many decades, unyielding to the challenges of concerned patrons, teachers, and government officials. "Out of date curricula, anachronistic teaching methods ...misallocation of public resources ...excessive centralization ...and neglect of teaching as a profession" (Otis, 1997 p. 34) all contribute to this overwhelming problem. One only has to look at the small contribution to the science community from Latin America (da Costa, 1995) to conclude that the preparation of students in science and of science teachers in these countries leaves much to be desired. This in and of itself is surprising since there is a long history in Latin America of science and technology linked with the Mayan civilization (Ayala, 1995).

In the United States, similar circumstances haunt science educators. It was only a little more than a decade ago that *A Nation At Risk* (Gardner, 1983) and other types of "evaluative" reports described the bleak and problematic nature of education in the United States. Indeed, US educators are also battling out-dated curricula, anachronistic teaching methods, a shortfall in public funding, and general neglect of the profession.

Other factors have also contributed to the present status of science education found in many North/South American schools. Many individuals still hold to antediluvian beliefs about the nature of science, which in turn affects the education of science teachers. Many teachers who are qualified to teach in the current systems are, for the most part, prepared in a perfunctory manner. A growing number enter the teaching profession without formal education in the field of science education, armed only with an emphasis on science content and laconic methods for the transfer of this content to their students. Not enough time is given to the pedagogy of teaching science, learning as it applies to the understanding of science, or current curricular reform in science education. Instead, emphasis is placed on the coverage of content through repetition and rote memorization as evidenced by current practice in schools in each of the participating countries in this report. The lack of pedagogical content knowledge only reinforces teachers' tendencies to stick closely to notes that they have prepared and avoid dealing with student questioning, which could unveil the teachers' lack of understanding of science (Stoll, 1993).

This is disturbing since both UNESCO's declaration for Project 2000+ (1993), and the American Association for the Advancement of Science's

Project 2061 (1989) indicate that sound science education is fundamental to strengthening scientific and technological literacy, which in turn is essential for acting responsible and sustaining development within any country. As Holbrook (1993) clarified, "Project 2000+...[is] a scientific and technological dimension of basic education in the context of education for all" (p. 4) that must occur for developing countries to reach necessary levels of literacy.

Reform in science teacher education, however, will not be at a small cost. Consideration will need to be made for the diversity of students and conditions within each country. Programs for science teachers will need to be developed keeping in mind the cultural beliefs and practices of their future students. Without the development of teacher preparation programs that take into consideration the level and diversity of student populations, that reflect goals of scientific and technological literacy, and that consider how these important facets of reform come together, it will be difficult to sustain change vis-à-vis science teacher education in any venue. The question arises as to how we go about making changes in preparing science teachers so that future teachers might address these critical issues. Individually, we have found that it is difficult at best to sustain rich dialogues and reflection that gets at the heart of the problems facing us all in the preparation of science teachers.

CHANGE AND SYSTEMIC REFORM

As concerned science educators, we began by discussing ways in which we might ameliorate key issues facing us. As our discussions began, independently we turned to the literature on partnerships, change and reform. We knew that incremental steps were better than wholesale changes (Aguerrondo, 1992). In light of that fact, we began by examining several key issues that would lead to partnerships that could assist in improving the preparation of science teachers. But as Crawley (1998) indicated, there are differences in the ways in which to define such alliances. "Networks" are one category of alliances in which exchanges of information are made between two or more parties who share similar interests, without any knowledge of whether there is perusal or processing of the information by the parties. A second category of alliances include "partnerships." Partnerships suggest an exchange of resources or tasks between dissimilar entities, a reciprocal arrangement in which learning may or may not occur. The idea of a partnership embodies alternative paradigms. It embraces individual and institutional renewal and offers greater probability that the agenda of research and the agenda of practice will be parallel and significant (Goodlad, 1988).

This suggests that strong relationships are planned, intentional, long term and are aimed at solving real problems (Shive, 1997). As part of the planning, it is key that a variety of stakeholders are included within the development of the planned partnership (Fullan, 1993). A critical component of this process is for the partners to give up their traditionally held myths about each other and learn to respect each other's strengths (Rudduck, 1991). Partners must share parity in the relationship. As well, there must be a clear purpose, vision, or goals for the partnership; and these types of partnerships should be developed from the bottom up (Shive, 1997).

With the above in mind, it seemed reasonable to start by examining models of change as a guide for the development of our partnerships. Although there are many effective models of change, most contain basic steps and considerations to use to restructure or reorganize. Although for each of our partnerships liberties were taken in applying a change model, we followed basic change principles and used these principles as guides in the development and implementation process (Cushman, 1993; Lieberman & Miller, 1990; Sizer, 1991; Timar, 1989). These principles included:

1. Agree on a new vision and from this develop the goals.
2. Get conflict in the open.
3. Understand that you cannot accomplish all your changes at one time. Change takes time.
4. Go beyond surface issues in order to promote quality and equality for all students.
5. Recognize the importance of building partnerships and networks with all involved and include them early on.

Change is never easy, and as Cushman (1993) lamented, "The first steps to change--agreeing that a problem exists, and setting goals to solve it--sound deceptively simple" (p. 4). Knowing that all must have their say and raise concerns about the visions projected for the partnerships, we spent over a full year developing goals for simultaneous renewal. More specifically, we attempted to come to a common goal of connecting university-to-university, faculty-to-faculty, teacher education program-to-teacher education program to examine the issues and set forth objectives that we all could implement to improve science teacher education in our respective countries. Some of these objectives included:

- To improve the quality of science teaching in public and private schools through effective science teacher education.
- To aid in the development of a philosophy of science teaching that is consistent with current research on science teaching and with the culture of the participants.

- To broaden and deepen the understanding of science teachers knowledge (in Bolivia, Chile, Venezuela, and the United States) of the current trends and research on science teaching.
- To develop and implement appropriate teaching strategies for teaching science in Bolivia, Chile, Venezuela, and the United States through simultaneous renewal.
- To develop curriculum that is consistent with teachers' philosophies and compatible with current trends in science teaching.
- To prepare a cadre of science educators who can act as leaders in their own country and cooperate amongst the partner countries in sustaining change.
- To develop long term and permanent opportunities for a cooperative research program that examines science teacher education across the globe.

We understand that the situation of various educational systems cannot be improved by isolated and simple proposals and strategies. Instead a deep transformation will provide solutions to the poor quality of education and provide for systemic change throughout the whole system. It is also apparent that in order for systemic change to occur, both the educational systems and science teacher education will need assistance.

TALES OF THREE COUNTRIES

Given the preceding discussion, the following represents the stories of the three countries (Bolivia, Chile, and Venezuela) involved in partnerships with counterparts in the United States. Each of these anecdotes is grounded in the context of the country, and includes particular issues related to the education of prospective science teachers. Each anecdote is unique in its resolve of problems found in a given country. We do not see these stories as prescriptive in nature. Rather, we see the narratives as descriptive, providing the reader an opportunity to gain some insights into the issues and history of science education in three Latin American countries and visions of how to ameliorate the critical issues surrounding science education through partnerships.

Bolivia

In any setting, teaching is not an easy task. Yet, teaching science in a developing country such as Bolivia is difficult since many science teachers operate with little or no equipment and very few supplies. Few Bolivian schools have funds for staff development or professional enrichment.

Hence, instruction in most Bolivian classrooms depends heavily on copying text material word-for-word along with rote memorization drills.

The current system of education in Bolivia provides for four levels of instruction: preschool (two years, pre-kindergarten to kindergarten), elementary school (five years, first through fifth grade), intermediate school (three years, sixth through eighth grade) and secondary school (four years, ninth through twelfth grade). Although the schooling maybe considered compulsory, many children attend sporadically or not at all. Little is done to ensure that all students attend school, especially in the rural settings of the country. Most all of the schools in Bolivia are state supported (88% of the enrolled student population) with far fewer attending private schools (12% of the enrolled student population), most of which are supported by the Catholic Church. It is also worth noting that the majority of the schools are found in the urban areas of Bolivia with few located in the rural settings. Most all of the rural areas in Bolivia either do not have schools, or if they do, children attend infrequently or drop out as early as the third grade to work in the fields. When schools are found in rural areas, they are ill equipped, lack essential amenities (electricity, running water) and have little in the way of curricular materials.

The Issues

Even though schooling through the secondary level is obligatory by law, the Bolivian population has a large percentage of illiteracy. This is largely due to the fact that many individuals in Bolivia are segregated in rural areas and small towns and still maintain ancient cultural values that do not place an emphasis on "modern education." For example, the indigenous populations (Quechua and Aymara) incorporate Catholicism and mysticism into a belief system that values traditional explanations for natural phenomena over school science explanations.

In regard to science education, it is fair to say that in general terms the quality of instruction offered to Bolivian students is low. Only the 12% or so who attend private schools have access to modern equipment, textbooks, technology, adequate physical plants, and other curricular resources. Unfortunately, private schools are very expensive and only the elite social classes of the Bolivian population can afford to send their children to these schools.

In addition to the aforementioned obstacles, there are other factors that contribute to the poor quality found in many Bolivian schools. First, as previously mentioned, the educational system is elitist and does not take into consideration the social, cultural, and ethnic diversity of its population. This becomes even more critical in light of the fact that many individuals still hold to mystic or ritualistic beliefs about the nature of science. They use traditional rituals to ward off evil spirits, control weather, produce good crops, and encourage conception. Such practices are often in direct conflict with school science. This is disturbing since UNESCO's declaration for Project 2000+ (1993) indicated that, "Sound education is fundamental to the strengthening of higher levels of education and of scientific and technological literacy...scientific literacy and technological literacy are essential for acting responsible and sustainable development" (p. 6). Without the development of programs that take into consideration the diversity of the population, meeting the goals of Project 2000+ in Bolivia will be difficult at best.

Closely related to the first concern, is the issue of language. The Bolivian educational system has ignored the fact that many individuals, especially in rural areas, do not speak Spanish as their first language or at all. Spanish, Quechua, and Aymara are all official languages in Bolivia, yet few teachers in the Bolivian system are bilingual. The language issue is compounded by requiring students learn yet another language in school, the language of science. Scientific terminology and the style of language used in science can cause difficulties for students as they attempt to learn science concepts (O'Toole, 1993) in their first language. Learning the language of science simultaneously with Spanish adds another layer of complexity for rural schools.

Finally, the teachers in Bolivian school systems are, for the most part, prepared in a perfunctory manner. Teachers are educated in a Normal School (a 2-year teacher training institution) with an emphasis on science content and superficial methods for the transfer of this content to their students. Little if any time is spent on the methods of teaching science, learning theory, or curricular theory. Most of these Normal Schools are poorly funded and lack the material resources (e.g., technology, curriculum, laboratory equipment) to provide teachers with appropriate experiences in preparation for teaching science. As previously stated, emphasis is placed on the coverage of content through repetition and rote memorization. In addition, many teachers in Bolivia who teach science have no formal training at all. These teachers, called "interim teachers," are hired because of teacher shortages due primarily to low pay. The teaching of science and math by teachers who are not qualified often leads to rigid tradition-bound teaching styles.

The Current Reform

Although the odds seem insurmountable, Bolivia is making headway regarding reform for science education and education in general. As a basis for the reform efforts in science, Bolivian educators agree that reform (or change) is never easy and the first step to implementing change is recognizing that problems exist (Cushman, 1993). In addition, it is understood that the actual situation of the Bolivian educational system cannot be improved by isolated and simple proposals and strategies, but by a deep transformation that provides for systemic change throughout the whole educational system. It is also apparent that in order for systemic change to occur, the school systems and teachers will need assistance. More specifically, Bolivian science educators will need an:

> Environment that is both enabling and motivating--providing sanction, protection, capacity, knowledge, resources, and the opportunity to change--combined with a set of expectations and the sensitivity to know when, where, in what direction, and how hard to push. (Donahoe, 1993, p. 302)

In light of the aforementioned needs, there are some basic designs for educational reform in Bolivia. A first step in the reform effort is to address the education of all students in Bolivia, both rural and urban. Consideration will need to be made for the diversity of students within the country itself. Curricula will need to be developed keeping in mind the cultural beliefs, rituals, and practices that are part of students' everyday life in the Bolivian campo (countryside). As Stoll (1993) so aptly explained, many of the textbooks and curricula in developing countries are North American in nature and do not take into consideration the cultural, ethnic, and linguistic diversity of a country such as Bolivia. The reform effort must focus on the development of new curricula that will utilize traditional technologies and crafts as part of the focus. In addition, an emphasis is being placed on the development and implementation of technology that will assist in the instructional process.

In addition to changes in curricula, teacher education itself is being radically changed throughout the country. Science teacher preparation programs will no longer exist solely in the Normal Schools. Universities are implementing programs that focus on providing contemporary, high quality preservice education programs that have as their focus not only the learning of content, but also, learning about learning. More specifically, the education of teachers will focus on the cognitive, affective, and physical development of students. Contemporary methods of teaching will also be taught to the preservice teachers, such as the whole language approach, constructivist methods of teaching, and cooperative learning. Emphasis will be placed on teaching children critical thinking and problem solving skills rather than the rote memorization of facts related to science.

The Project

The partnership in Bolivia is unique. Unlike many partnerships where there is a link with a single institution, this partnership focuses on a cadre of individuals from three institutions (Universidad Privada Boliviana, Cochabamba Cooperative School, and East Carolina University) who after a year of discussions, made a commitment to systemic change in Bolivia. Although we have established links to various institutions, each person in the group comes to it because of an interest in change in science teaching and not because of some predetermined top-down agreement. From the beginning, the focus has been on the improvement of science teaching in Bolivia.

Many discussions led us to conclude that the traditional routes of teacher preparation may not be the best place to impact and change science teacher education, since government control, bureaucracies, and local politics have a tight hold on how schools are operated. Instead, the partnerships began by focusing on the development of a cadre of experts who are Bolivian, then providing these individuals with the support to impact others. In addition, the partners considered how to link faculty and students from East Carolina University to the partnership to further their understanding of science teacher education. It is from this perspective that this group of individuals established the partnership and developed a plan of action.

The overarching goal of the current project is to prepare teachers (inservice or preservice at the secondary level) who can teach students to be scientific and technologically literate life long learners. In addition, substantive efforts are being made to educate teachers to the needs of the students who are non-Spanish speakers or at best, speak Spanish as a second language.

In addition to the above, plans are being made to implement the following:

- Ongoing in-depth staff development programs for practicing teachers.
- Summer and winter institutes for continued development of teachers' knowledge base.
- Advanced degrees in the area of teacher education.
- Research programs that examine critical issues in education pertaining to Bolivia.

– Assessment strategies that are more closely tied to the current changes in curriculum and instruction.

Because of the extreme difference between existing and proposed teacher programs, and understanding that change cannot occur rapidly, a 4-year plan was developed. Each year, a single change activity, addressing one or more goals will occur. In each successive year, the project will implement an additional change activity to address remaining goals. The important component of the plan has been to go slowly, to be flexible, and not demand that all of the project goals be completed in three years. We also developed consensus that if the changes were not positive, or if changes were being made too fast, or were not meaningful, then the action plan could be altered.

The Action Plan

Year One: Establish an Instructional Enrichment Institute

The partners in the project will establish an annual winter (US summer) Instructional Enrichment Institute to be held on the Cochabamba Cooperative School Campus. This will be comprised of a series of professional workshops on classroom instructional theory and methodology to be presented by faculty and/or associates of East Carolina University (ECU). Those inservice and preservice teachers who participate in this program will act as trainers in the process of facilitating current methods and theories of sound educational practices for other Bolivian teachers. It is important to note that the methods and theories presented will be based on current knowledge of teaching that reflects the reality of Bolivian culture and values. Therefore, Bolivian educators will be working closely with ECU faculty to establish the curriculum of the proposed program.

Year Two: Establish a Post-Graduate Degree Program

Until recently, 4-year undergraduate programs for the preparation of science teachers were not even considered in Bolivia. Most teachers were 'trained' as part of a 2-year Normal School preparation. In the last four years, a transition has been made allowing future secondary science teachers to receive a 4-year degree in preparation for teaching science. Yet, to fully understand the nature of teacher preparation, one has to realize that institutions of higher learning are ill equipped to address the needs of future teachers. And, in many cases, the political infrastructure of the institution will not allow for the implementation of a more rigorous study of science pedagogy. With that in mind, the partners in Bolivia discussed the matter and concluded that a focus on educating teachers at a master's degree level would do more for science teacher education than focusing on undergraduate degree programs. It was with this in mind that plans are being made to develop and implement a master's degree at the Bolivian Private University (Universidad Privada Boliviana, UPB) for the purpose of encouraging Bolivian teachers to seek and secure a post-graduate degree in the field of science education. The establishment of the program will be a cooperative effort between the parties involved in the partnership. Courses for this master's degree will be taught at UPB for credit by UPB faculty, Cochabamba Cooperative School faculty, and potentially faculty from the East Carolina University. Additionally, teachers who participate in the Instructional Enrichment Institutes will receive university credit toward their master's degree at UPB. Only Bolivian university graduates with degrees from a Normal or an accredited 4-year degree program and who possess Titvio Provision National (Bolivian teaching certification) will be eligible for admission to the masters program.

Year Three: Establish the Continuing Education Center (CEC)

Cochabamba Cooperative School and the UPB have provided land for the construction of a resource facility providing research, training, and development opportunities for Bolivian science teachers, Cochabamba Cooperative School faculty, and faculty from East Carolina University. The Continuing Education Center (CEC) would be a complete instructional resource facility. It would offer the facilities and resources necessary to provide technical and professional assistance to Institute members for improving the quality of instruction in their individual classrooms.

The Center would provide a resource facility for science teachers in Bolivia to visit and from which to learn. This is a dire necessity as teachers in the Bolivian public school system lack the monetary resources to pursue their own professional growth. The CEC will provide teachers with the opportunity to: (a) study and/or pursue research at the professional/classroom library; (b) observe instructional techniques in demonstration classrooms; (c) create their own teaching models and instructional materials in the resource workroom; and (d) learn from the most advanced technologically equipped educational computer facility in Bolivia.

Chile

Systematic teacher training began in Chile in 1842, with the foundation of an "Escuela Normal" (Normal School) where all persons who wanted to become teachers for primary schools received their training. In 1889, the "Instituto Pedagagico" (Pedagogical Institute) where teachers for secondary schools were trained, was founded. It was later annexed to the Universidad de Chile. Finally, in 1950, at the Universidad de Chile, the "School for Early Childhood Educators" was created. It was in this institution that kindergarten and pre-primary school teachers were trained.

Since 1974, the preparation of all teachers has been the responsibility of the universities, but in 1980 a law was issued that also allowed the "Institutos de Educacion Superior" (Institutes of Higher Education) to undertake teacher preparation. In that year, a total of approximately 8,000 vacancies was offered by all institutions of higher education for the teaching profession and more than 30,000 students were being prepared to become teachers (Rioseco & Roro, 1984).

Although the Ministry of Education has traditionally controlled all teacher training throughout the country, each institution for higher education has its own curriculum for preparing teachers. The preparation of science teachers is no different from the training of other secondary teachers. It takes five years, with the last year for students to complete a practicum in schools and write a thesis, which is required to graduate as a teacher. In most institutions, the curriculum is organized in two big blocks, the general pedagogy training and the specific training for one or two subjects. Thus, teachers can be prepared to teach math, math and physics, physics, math and computation, chemistry, biology, biology and chemistry, natural sciences and physics, natural sciences and chemistry, or natural sciences and biology. In the following description, "science" will include natural sciences as well as mathematics.

In some cases, science subjects are taught during the beginning years of the preparation program and the pedagogy courses are taken in the last two years. In this model, teachers are prepared in two cycles: a basic cycle, whose curriculum is developed along six semesters of sciences at the university, and a terminal cycle, developed along three semesters in the area of education. The students who have finished the basic cycle can choose to become teachers or to attend another terminal cycle at the same school/college where they finished their basic training. The additional five semesters leads to a degree equivalent to a Masters' Degree in Science. Similar to this model is "Bachillerato". This cycle is intended to orient students and reinforce their science training, before they make their professional choice. After obtaining a bachelor's degree, they can decide to become a science teacher or continue their science training in any other school at the university.

These models have a fundamental problem. Basically, students see themselves as scientists and not as teachers. To compensate for this problem, another model has been used where science courses are placed in the first four years of the student's study along with pedagogy courses.

During the first year, there are four or five science courses and only one course in pedagogy. This pattern changes during the program: the number of pedagogy courses increases until the fourth year where the curriculum includes four pedagogical courses. In this way, the number of pedagogy courses increases gradually, while the number of science courses decreases gradually during the same period of time. This means that at the beginning of their program, students receive a more intensive science training and at the end, more intensive educational training.

The science courses attempt to cover all science content that teachers are to teach in secondary schools. Pedagogy courses include Fundamentals of Education, Educational Psychology, Curriculum, Evaluation, Teaching Methods or Didactics, Counseling, and Educational Administration, among others. In most cases, credits corresponding to science courses amount to 60% or more of their program, while credits corresponding to pedagogy courses amount to 40% or less.

The Current Reform

A new plan for teacher preparation is now being proposed to the Ministry of Education that affects primary and secondary school programs. The main change moves from a subject-oriented approach to an integrated approach, where subjects should disappear, giving place to pedagogy blocks that will integrate curriculum, evaluation, and teaching methods along four semesters. Another feature of the plan is the early inclusion of students in the school system, giving them the opportunity to practice in front of real pupils and experience real school environments from the very beginning of their preparation. Under this plan the emphasis will be placed on practice rather than just theory.

Courses like Education Fundamentals, Educational Management, Psychology, and Counseling will also incorporate visits to schools and discussion and analysis of real problems as part of their methodology and assessment. This plan was implemented in 1998. All education faculties (at all Chilean Institutions) are proposing their innovative curriculum for teacher preparation to the Ministry of Education, and will receive funding when the plan is officially approved.

The Project

Unlike the partnership in Bolivia, this partnership focuses more on the development of programs at the undergraduate or preservice level. Both the Universidad de Concepcion (UC) and East Carolina University (ECU) are undergoing renewal efforts to strengthen their preservice programs in the area of science. This joint UC and ECU program is designed to increase the educational potential in science for students at all educational levels and to improve the teaching and learning processes at both universities. It is proposed that this be accomplished in part by conducting research in science education which focuses on preparing qualified professionals to teach science in secondary school settings.

As part of this link, joint efforts are being made to develop curricula for science education as well as apply research outcomes. In the partnership, we feel that it is critical that we establish leadership and promote the creation of working groups that can contribute to the progress of science education at both institutions. As a result of these collective voices, faculty will develop joint science education research projects in areas such as:

- Science education in its social context
- Learning of science
- Issues in the historical interpretation of science
- Teaching of intellectual skills
- Science teaching as discourse
- Linking industry and science teaching
- Science teacher preparation and professional development.

This will lead us all to a greater understanding of science teacher education within our own contexts and in the broader context of the partnership. Currently, efforts are being made to co-develop courses that could be offered via Cu-See-Me technology. As part of the initial agreement, both campuses were provided with this technology and links have been established to incorporate a broader view of the world, and hence science, in all of our curricula.

It has been apparent from our discussions that a significant pitfall of both institutions is the virtual isolation based on geographic location. Both the Universidad de Concepcion and East Carolina University are located away from other major metropolitan areas of the state or country. Both institutions are near oceans and have economies that are influenced heavily by forestry, fishing, and agriculture. We have established exchanges as a way of breaking down the walls of isolation. We have proposed in our project to establish exchanges that will:

- Share information about the activities and research work pertaining to courses taught at each institution;
- Share scientific publications, research materials and teaching materials;
- Share information pertaining to the development of courses for teacher preparation;
- Share members of each institution for the purpose of collaborating in the collaborative teaching and research.

Venezuela

The Venezuelan educational system was described by Dr. Antonio Luis Ordenas, the former Minister of Education, as "a fraud," thereby admitting the failure of national policy (personal communication, November 15, 1995). In spite of this criticism, not much has been done to remedy the evils of partisan politics and political favoritism that seem to control educational reform in Venezuela. At the same time, official reports show that the percentage of the nation's population living in poverty and extreme poverty has grown to 83%.

Under these conditions, it is impossible to develop a teaching-learning process capable of guaranteeing a complete education and vindicating democracy as a form of government. While it is undeniable that progress has been made to find solutions to the crisis, the observable results do not meet society's expectations, do not match electoral promises, and are out of proportion to the amount of money presently invested by the Venezuelan government. At the time of this publication, the Ministry of Education is incapable of providing the necessary leadership at a national level, decentralization remains an empty word, and party politics are still the determining factor in educational decisions. School buildings need urgent repairs and teaching materials are either nonexistent or inadequate. Inservice education of teachers is systematic and has no significant influence on curriculum development. And lastly, teacher preparation is out of touch with educational reality; university candidates for teacher preparation programs are few in proportion to the real needs.

As if this was not enough, another detrimental factor has surfaced among Venezuelan youth: the loss of traditional values. Adolescents, in an effort to look and act westernized, are giving up many traditional values. This condition has been strengthened by the widespread social disillusionment and sped up by means of mass media and more recently, the Internet. The scarcity of job opportunities for the group that has been excluded from, or has dropped out of, the educational system also impedes the growth of a society and widens the social gap.

The Educational System

There are nine years of compulsory basic education (which does not include high school) in Venezuela, reaching 92% of the school-age population between 6 and 15 years old. At this level there are serious problems in the curriculum, especially in the areas of reading and writing, logical reasoning, mathematics, and awareness of national identity. However, the most serious problem, because of its social repercussions, is the high dropout rate and the difficulty of attracting dropouts back into the system. There is also a high incidence of repeating courses, as well as a problem of bridging the gap with the next level of high school education. The current curricular reform put forth by the Ministry of Education is an attempt to deal directly with the problem of high school articulation.

Thirty-three percent of the population goes to high school. The high schools suffer from low academic standards and lack of future employment relevance. The high schools also suffer from a shortage of properly prepared teachers, inadequately equipped laboratories, and out of date technology. Consequently, the present professional education systems fail to meet the needs of the industrial sector, or the expectations of society. To make a bad situation worse, high school is still the traditional funnel through which the overwhelming majority of students go on to higher education. High schools thus focus on college preparation instead of the numerous job-training programs that could be promoted at this level, and that would be more in touch with the service sector labor-market needs of small and medium industry.

Lastly, a high school graduation certificate does not automatically guarantee entrance to public universities. As a result, Venezuela has seen a proliferation of private universities and academies, most of which have low professional standards and whose primary interest is to make a profit and take advantage of the large number of students who want a university degree.

Higher Education

Many Venezuelans believe that Venezuelan higher education has lost its sense of mission for the future. It appears unable to keep in focus its policy of developing human resources, of keeping its academic and administrative programs up-to-date, or of keeping in touch with the real needs of the country in order to fulfill the vital role that has been entrusted to it. The traditional style (i.e., top down decision making) of university administration still persists and regrettably mirrors national governmental procedures. The lack of coordination between the planning departments and the ruling bodies of the universities, and the absence of any efficient evaluation system, is evidence of the crisis in higher education. But at the same time it should not be forgotten that 93% of research in the country is carried out in these public universities, in spite of financial cutbacks.

The political, social, and economic structure of Venezuela is at present in the midst of its worst crisis in recent times. It is obvious that the model based on oil revenue is no longer valid. The good news is that plans are underway to weaken the power of big government. The recent election promises to privatize the basic national industries, attract transitional capital, negotiate new loans with the international banking system, modernize state institutions, and open Venezuela to the world market with the globalization of the economy. With this positive outlook in mind, institutions have formed partnerships and made much needed progress.

Reform Partnerships

During the early 1990's Dr. John Penick, then from the University of Iowa, developed a working relationship with the science faculty of University de los Andes in Merida, Venezuela. Their first task was to establish a new science education center and procure funding. Dr. Mauro Briceno, from Merida, spent the 1994-95 academic year at the University of Iowa where details concerning mission and goals of the center were established. The mission of the Science Education Center of Universidad de Andes states that it will be dedicated to the promotion and improvement of the teaching and learning of science, mathematics, and technology (SM&T) of all persons at all levels throughout Latin America and the Caribbean. In addition the Center set out to fulfill a variety of functions including:

1. Designing and conducting research related to the teaching and learning of SM&T at all levels.
2. Designing and delivering preservice and inservice teacher education, including methods of teaching, the content to be taught, and new technologies.
3. Providing consultation, leadership and training in the development, and instruction for education in SM&T.
4. Developing appropriate assessment techniques.
5. Developing policy statements related to the teaching and learning of SM&T.
6. Establishing and maintaining cooperative links with science education centers, science teacher associations, established groups of SM&T researchers, and science teacher educators worldwide.
7. Establishing and housing a library for materials and educational technology useful for SM&T education.
8. Disseminating information and ideas related to SM&T worldwide.
9. Creating and coordinating SM&T activities for pre-university students.
10. Encouraging and development of SM&T professional associations.

By 1996 the Center had been funded, and set forth to carry out its mission. At this point, The University of Nebraska was invited to join this partnership, resulting in several visits and numerous new initiatives. One of these initiatives was directed at establishing a working relationship between the University Science Education Center and the new government-funded regional inservice facility.

The Regional Center of Support for Teachers

The state of Merida Regional Center of Support for Teachers (CRUM) was established as a model of professional development for teachers. CRUM is located near the town of San Juan de Lagunillas on a large hacienda about 30 minutes from the Science Education Center. The facility and grounds have been designed to dignify the profession of teaching and have placed the State of Merida and the new Science Education Center in the international spotlight with this inservice facility.

The hacienda can house and feed 100 inservice teachers at a time and includes: a computer lab, four classrooms including a science laboratory, a modern reference library complete with on-line search capability, and a large outdoor nature center for environmental studies. Four groups of 25 teachers each are sent to the CRUM for week long workshops. One of the unique features of these workshops is the fact that participants are required to again return several months later for a follow-up workshop to assess progress and help with any implementation problems they may have encountered. The University of Nebraska has been helping identify both areas of need for these inservice programs and presenters from around the world who might assist in delivery. For example, many of the *Project Wild* (Western Regional Environmental Education Council, Inc., 1992) activities have recently been translated into Spanish by a group in Puerto Rico and are ready for presentation throughout Latin America.

The ongoing challenge will be the continued cooperation between agencies. An unexpected role of communication and facilitation by the University of Nebraska has emerged. For example, while the new Science Education Center, which is housed within the science department, included preservice science preparation as a goal, it had virtually no contact with the education faculty who were responsible for preservice education. Likewise, the Science Education Center had not developed a plan for regional inservice with the new CRUM. The international partnership brought to light these fragmented efforts and opened lines of communication. In addition to internal communication enhancement, the partnership has opened the door to greater external communication.

Faculty Visits

As a result of trips to Merida from North American faculty, arrangements have been made for Venezuelan physics professor, Dr. Alejandro Noguera, to visit several universities in the United States. The primary goal is the translation of a hands-on, process- oriented undergraduate physics program created by Dr. Robert Fuller at the University of Nebraska. Both the reform-oriented concept and curriculum materials will thus be transported to Venezuela. Additional projects have been designed and await funding agency approval.

CONCLUSIONS

Although there may be many reasons to doubt the changes suggested in these countries will occur, the vision has been established. Without a vision of what can be, we will never start our journey. And, it is through this journey that we encounter difficulties that provide us with experiences, and experiences in turn bring us wisdom to understand change. Ultimately the success or failure of projects such as these rests with the individuals who have a stake in the results of the change. Many of us have good ideas and good intentions of sharing our ideas with others. Yet, we must ask ourselves if our vision is shared by all of those individuals whom it will impact. Can we be sure that those who are most affected by change see a need to change? Or, do they have the attitude, "If it is not broken, why fix it?" Without the commitment of the all involved in the project, the vision of one becomes little more than an imposition on others.

With this in mind, the partnerships established are a big step for the educational systems in developed and developing countries as we move toward systemic reform. Overhauling the science teacher education process, developing new curricula sensitive to the nature and needs of the students, building scientific literacy, developing problem solving and decision making skills as a critical component of students education, as well as empowering students to be life long learners are all worthy goals that will take time and substantial effort to implement. However, taken together these changes could have an immense impact on science education in all participating countries.

It is clear to us that all of our institutions are undergoing similar types of reform efforts. Studying and sharing information about how we address these systemic reforms will only strengthen our own science teacher education programs. As far as the impact for institutions, faculty, and students in the United States, our overall goal has been to enable faculty and students to develop a global view of science education research and teaching through our collaboration. It is through these collaborative partnerships that immense educational problems of South and North America become clear. But with each of these challenges comes an assortment of opportunities for symbiotic relationships and personal and professional growth. As an example, we all are aware of the growing need for bilingual teachers in the U.S. and the need for greater multicultural awareness throughout our programs. There is no better way to gain these badly needed insights than by forming partnerships with our international colleagues.

REFERENCES

American Association for the Advancement of Science (1989). *Science for all Americans.* New York: Oxford University Press.

Aguerrondo, I. (1992). Educational reform in Latin America: A survey of four decades. *Prospects, 22*(3), 353-365.

Ayala, F.J. (1995). Science in Latin America. *Science, 267*(5199), 826-827.

Crawley, F. (1998, April). *Transforming science teaching through action research: Professional communities as venues for rethinking research, practice, and policy.* A paper presented at the Annual Meeting of the National Association for Research in Science Teaching, San Diego, CA.

Cushman, K. (1993). So now what? Managing the change process. *Horace, 9*(3),1-11.

da Costa, L.N. (1995). Future of science in Latin America. *Science. 267*(5199), 826-828.

Donahoe, T. (1993). Finding the way: Structure, time, and culture in school improvement. *Phi Delta Kappan, 75*(4), 298-305.

Fullan, M. (1993). *Change forces: Probing the depths of education reform.* Bristol, PA: Falmer Press.

Gardner, D. (1983). *A nation at risk: The imperative for educational reform.* Washington D.C.: U.S. Office of Education.

Goodlad, J. (1988). School-university partnerships for educational renewal: Rationale and concepts. In K. Sirotnik & J. Goodlad (Eds.). *School-university partnerships in action: Concepts, cases and concerns* (pp. 3-31). New York: Teachers College Press.

Holbrook, J.B. (1993). Basic and applied science education in developing countries: Trends and needs of the 21st century. *Science Education International, 4*(1), 3-10.

Lieberman, A., & Miller, L. (1990). Restructuring schools: What matters and what works. *Phi Delta Kappan, 71*(10), 759-764.

Otis, J. (1997). Education crisis. *Latin Trade, 5*(8), 32-37.

O'Toole, M. (1993). Science, technology and communication: Utilization depends on access. *Science Education International, 4*(4), 21-24.

Rioseco, G., & Roro, I. E. (1984, April). *The present situation of science and technology education in Chile: National Study.* A paper presented at the International Symposium "Interests in Science and Technology Education," Kiel, Germany.

Rudduck, J. (1991). *Innovation and change*, Milton Keynes, UK: Open University Press.

Sizer, R. T. (1991). No pain, no gain. *Educational Leadership, 48*(8), 32-34.

Shive. J. (1997). Collaboration between K-12 schools and universities. In N. E. Hoffman, W. M. Reed, & G. S. Rosenbluth (Eds.). *Lessons from restructuring experiences: Stories of change in professional development schools* (pp. 33-50). New York: SUNY Press.

Stoll, C.J. (1993). Science education in developing countries: What's the point. *Science Education International, 4*(4), 13-16.

Timar, T. (1989). The politics of school restructuring. *Phi Delta Kappa, 71*(4), 265-275.

UNESCO. (1993). International forum: Scientific and technological literacy for all declaration. *Science Education International*, *4*(3), 6-7.

Western Regional Environmental Education Council, Inc. (1992). *Project WILD*. Bethesda, MD: Author.

Chapter 11

International Science Educators' Perceptions of Scientific Literacy

Implications for Science Teacher Education

Deborah J. Tippins[1], Sharon E. Nichols[2], and Lynn A. Bryan[1]
with Bah Amadou[3], Sajin Chun[4], Hideo Ikeda[5], Elizabeth McKinley[6], Lesley Parker[7], and Lilia Reyes Herrera[8]
[1]University of Georgia, U.S.A.: [2]Eastern Carolina University, U.S.A.: [3]Universite de Conakry, Guinea: [4]South Korea: [5]Hiroshima University, Japan: [6]University of Waikato, New Zealand: [7]Curtin University of Technology, Western Australia: [8]Universidad Pedagogica Nacional, Colombia

Abstract: Science educators worldwide are calling for the development of scientific literacy in today's schools, yet there is little consensus as to what criteria or goals might constitute the attainment of scientific literacy. In this chapter, we explore the diverse meanings international science teacher educators have for scientific literacy as it relates to their own cultural backgrounds and professional practices. We conducted the study in the interest of preserving two types of context: the unique context of a science educator's life story and the biographical contexts that enrich the meaning of the individuals' perceptions of scientific literacy. Participants involved in the study included six science teacher educators representing: Guinea; West Africa; South Korea; Japan; New Zealand; Austria; and Colombia. We initiated interviews with participants using several open-ended questions with the intent to elicit conversational responses. We wrote the narratives presented in the study to preserve the insights shared by participants from their unique perspectives, and to avoid imposing an interpretation drawn from our worldview. Ultimately, the chapter highlights the ways in which scientific literacy is reflective of social, cultural and political situations that shape local communities and science teacher education practices.

With the advance of modern communications technology, the science education "community" is becoming a smaller place. Science educators are empowered by increased opportunities to share information/scholarship and ask questions that transcend national boundaries. In the midst of this media-propelled landscape, we have observed that science educators throughout the

S.K. Abell (ed.), Science Teacher Education, 193–221.

world have a great deal in common. In spite of vast differences in the cultures and politics of our homelands, we ask similar research questions and share concern for educational issues of importance to the global community. We are convinced of the necessity to create critical partnerships of mutual conversation with international science educators. Our observations and reflections have led us to conduct a series of interviews with science educators around the world and to explore and learn about their perceptions of scientific literacy.

The rhetoric of modern day reform is replete with references to scientific literacy and descriptions of how it might be "developed" or "obtained." An impressive number of science education researchers, reformers, and practitioners throughout the world have joined in the call for scientific literacy in today's schools. At the same time, others have expressed skepticism toward the goal of making the majority of people scientifically literate. Shamos (1995) described this goal as a "myth" and questioned the extent to which scientific literacy could be developed with a reasonable amount of effort. Reflecting on these criticisms, we wanted to interview science educators around the world and learn about their perceptions of scientific literacy. Because of their involvement with pre- and inservice science teacher education in their respective countries, we assumed that our interviewees would have some interesting perceptions concerning the goal of scientific literacy. We wanted to know how these scholars viewed scientific literacy in relation to their own cultural backgrounds and what the phrase meant to them on a personal level.

The science educators we interviewed are not a demographically representative sample of international scholars in the field. They have diverse roles within the science education communities of their homelands, and their science content specializations are varied. Some are in the beginning stages of their academic careers, some are broadening their educational background in science education, and others have long established careers in science education at their respective institutions. However, all of our interviewees have a strong interest in and commitment to pre- and inservice science teacher education. Their representativeness is irrelevant to the approach we took, since our objectives and assumptions are inconsistent with a generalized portrait of international science educators.

Because we were interested in eliciting personal perceptions of scientific literacy from each scholar/science educator, we began by asking each to consider whether he/she was scientifically literate. We did not assume an a priori definition of scientific literacy; rather, we wanted to gain insights from the biographical narratives of each individual. The interview protocol used in the study was modified from questions developed for an earlier pilot study of science educators' perceptions of scientific literacy (Tippins et al., 1998). Questions that were used to elicit responses included:

1. Do you consider yourself to be scientifically literate? Why or why not?
2. When you hear the term "science for all" what does this mean to you? What are the positive or negative aspects of the "science for all" notion?
3. Is there a connection between science for all and scientific literacy? Please explain.
4. How do you perceive your role as a science educator in the development of scientific literacy? What do you think should be the role of schools in the development of scientific literacy? What do you think should be the role of individuals, institutions, and others in the development of scientific literacy?

OUR OBJECTIVES

In this study, we were interested in preserving two types of context--the unique context of a science educator's life story and the biographical contexts that enrich the meaning of individuals' perceptions of scientific literacy. These objectives stand in stark contrast to the objectives that inform extant studies of professional roles such as the professoriate. These studies, although occasionally highlighted with brief snippets of verbatim dialogue, were purposefully designed to reflect generalized views (Boyer, 1990; Clark, 1987; Ducharme, 1993; Wisniewski & Ducharme, 1989). Where extant studies typically obscure the discursive nature of interviews (Ducharme, 1993) our intent was to preserve the "internal history of the developing discourse" (Mishler, 1986, p. 38) by representing the interviews in their near entireties. While our interviews were informed by guiding questions, we did not remove comments that were tangential to the topics of scientific literacy and science teacher education. Since our interviewees often related stories or described experiences that were an extension of their professional roles as science educators, we chose to include those perceptions, because "they enrich understanding and are integral to the personalities of these individuals" (Kagan & Tippins, 1995). In general, our editorial work was relatively minor and informed by member checks and collaborative dialogue with the interviewees. A combination of face to face interviews and e-mail conversations was used to facilitate the narrative process. For example, initial interviews with Hideo Ikeda and Lilia Reyes were initiated through e-mail. A subsequent trip to Japan enabled us to continue the conversation and member check with Hideo; and similarly we met with Lilia Reyes at the at the annual meeting of the National Association for Research in Science Teaching. Ultimately, we hope our readers will come to know these six individuals, gain insights about science education in their particular contexts, and become familiar with their highly personal perceptions of scientific literacy.

INTERNATIONAL SCIENCE EDUCATORS' PERCEPTIONS OF SCIENTIFIC LITERACY

Bah Amadou: Guinea, West Africa

Bah was born in Telimele in the northwest of Guinea and moved to the capital city of Conakry where he has spent most of his life. Initially, his academic goal was to be a chemical engineer. However, he soon found too many engineers in a country characterized by a basic lack of industry. Thus he started teaching and received a doctoral degree in chemistry. As he became more interested in teaching, he decided to enroll in a science education master's degree program in the United States. Having recently completed this program, he has returned to teach at the Universite de Conakry in Guinea. Bah described the reasons for shifting his career focus to secondary science education as a "synergy of many factors." He explained:

There is a general lack of support for scientific research in chemistry in my country. A lack of equipment and reactants made it difficult to publish in my area. At the same time, many higher education institutions were being created, and they needed teachers. Actually, they initially wanted to send me into the interior of the country to teach. But they forget about you--you get lost--so I chose to remain in Conakry, which is the capital.

Bah started his teaching career at the secondary level (Lycee), teaching grades 7-12 general chemistry. After teaching secondary science for several years, he moved to higher education. When asked to describe the nature of science teacher education in Guinea, Bah explained:

We don't have specialists in science teacher education. So any time we have an educational reform or problems, we have to call on specialists from other countries. We frequently call on science educators from Canada with whom we have a long-time cooperative relationship.

In general, Bah described science teaching in Guinea as very didactic in nature, characterized by a lack of hands-on experimentation or collaborative group work. He noted that teachers typically have strong science content backgrounds but very little experience with pedagogy. Teachers wishing to include cooperative and collaborative grouping strategies in the science classroom must first obtain permission from a supervisor.

Bah went on to describe some of the fundamental differences between rural and urban education in Guinea, saying:

> *In the capital, we can't have fewer than 50 students in the classroom. Classes are overcrowded and suffer from a lack of space. The opposite situation is true in the interior. Actually, it's funny--they reserve sites to build schools and people come along and build businesses on these sites. In the interior people were reluctant to send their kids to school. In rural areas some people were very conservative and fought progress. Although people are obligated to send their children to school, no measures are taken to enforce this--so it's a contradiction. Authorities do not anticipate the needs of a growing population. Things are changing, though, and presently the population in rural areas build schools and are provided with teachers.*

Bah emphasized that the importance of understanding different cultural groups cannot be overlooked in making sense of goals associated with scientific literacy in Guinea:

> *Each group has its own culture, and scientific literacy depends on how you perceive the world. Sometimes we try to understand from our own perspective or scale instead of looking from the inside out.*

While Bah felt that he was scientifically literate because of his background in the sciences, he struggled with the "vagueness" of the term. In general, though, he cited motivation, access to various forms of knowledge, and opportunities to be confronted with a problem as important components of scientific literacy. He described scientific literacy as "a very broad term that doesn't represent any content" and went on to qualify this by saying:

> *In general terms, I consider it as the ability to manage scientific material you haven't encountered before. Being aware of news and discoveries and understanding how you can assimilate all of this stuff...there's both a practical and a technical type of knowledge. Actually, the limits are fuzzy. But it seems like the person who is scientifically literate should also have some specific, limited knowledge. Not only can he fix a camera, he also understands the physics of a camera.*

Bah explained that the Guinea system of education emphasizes general literacy as characterized by the ability to read and write using the Latin alphabet. He was equivocal about such narrow parameters to describe literacy:

My ethnic group includes people who are very literate when it comes to using Arabic letters. We read the Koran, we write using Arabic letters. Although people typically don't use the Latin alphabet, don't speak or understand French, and don't have a formal education, they are very literate in areas of theology. Many of the authorities consider them to be illiterate, but I personally don't consider this population illiterate. Anyway, illiteracy is very high even in developed countries.

Bah paused for a moment and, on a reflective note, related a story that illustrated the dilemma:

I remember in 1961 there was a kind of eclipse--lunar eclipse. My father at that time had a radio. He knew the day before that there would be an eclipse. But most of the people did not have radios. My father told them that there would be an eclipse and that it was only a natural phenomenon that would last 10 minutes. But most didn't believe him, and when it happened, they were fearful and wailing. If you don't have access to information or an understanding of that information you will not be as literate. When people are superstitious they link it to political and social events and natural phenomena.

Like educators such as Morris Shamos in *The Myth of Scientific Literacy* (1995) or philosophers such as Henry Adams in *The Education of Henry Adams* (1918), Bah believes that it is unrealistic to educate all people in terms of scientific literacy. In explaining this belief, he emphasized:

You can't anchor the idea [of scientific literacy] to all people at the same time. Scientific literacy comes at the moment of wanting to learn. I think it's a personal decision--no one can make someone literate.

When comparing science education in Guinea and the United States, Bah made some distinctions between the two societies:

In Guinea you can function as an individual in society without having a high degree of scientific literacy, but this is not so in the U.S. I think it is because of the way that technology permeates everyday life. Technology and scientific literacy are closely connected.

Bah, in describing what he perceived to be a fundamental difference in education in the two countries, went on to say that:

I perceive a future generation in the U.S. that will no longer be able to write by hand. Technology will be taken for granted. The complexity of technology is so high in this country [the U.S.] that it is driving the future. This is not the case in my country. We have book-related knowledge, but we have limited access to other forms of knowledge. And we also gave more stratification--some live very well while many live in poverty.

Bah described the frustration he personally experienced when first faced with using technology as simple as the telephone:

I remember when I first got here to study, I made a call on the phone. But there was a machine talking and so I just stopped talking.

As Bah pondered the future and his possible role in facilitating the development of scientific literacy in Guinea, he was somewhat pessimistic:

Getting a position in my country is not easy. I am well qualified, but you need more than skill--you need connections.

At the same time, he shared with us his dream for the future:

What I would really like to have is a science learning center for teachers, but it requires much money, so I am skeptical. The center would be a focal point for collaborating with schools. The center could serve as a museum, for classes, promote new pedagogies, new approaches to preparing science teachers. Teachers could learn about other forms of knowledge and less authoritarian forms of teaching. In helping to develop scientific literacy the center would be an intermediary solution to a long term problem.

Sajin Chun: South Korea

Sajin was born in Ulsan, a fairly large city in the southeastern part of the Republic of South Korea. She noted that Ulsan was a small place when she was growing up, but is much larger today due to the influence of the automobile industry. Sajin was educated in Seoul, the capital and largest city in South Korea. After graduating from secondary school, Sajin attended Seoul National University and received a bachelor's and master's degree in Earth Science Education. She is currently in her third year of doctoral studies in science education at the University of Georgia. Although uncertain about the future, Sajin hopes to return to South Korea and become active in science education at the university level.

Sajin's first teaching position was in a middle school where she taught 2nd grade (8th grade in the U.S.) physics and chemistry. She eventually applied for admission to the doctoral program in science education at Seoul National University. During this time she taught Earth Science to prospective elementary teachers in the Inchon National Teachers College. Sajin described this period as "the best opportunity to really get me thinking about how to prepare teachers." After completing a year of her doctoral studies, Sajin made the decision to complete her degree in the U.S. Her decision was based on both professional and personal factors:

> *In our department there are five faculty members who majored in different subjects of science: astronomy, geology, geophysics, meteorology and oceanography. Though these professors are in the Earth Science Education Department for preparing middle and secondary grade teachers, they are more close to scientists rather than science teacher educators. So I could develop scientific knowledge deeply, but I did not consider pedagogy or pedagogical content knowledge seriously at that time. I think it is changing now, and other departments--Biology and Physics--already have several faculty who majored in science education, not pure science.*

Sajin's decision was also based on the desire to follow her husband to the United States where he had decided to earn a Ph.D. in Physics.

We were interested in Sajin's thoughts about science education in South Korea. When asked to describe the contexts with which she was familiar, Sajin responded:

> *In the classroom there are over 50 students. In the U.S., teachers have their own classroom, but in Korea it is reversed. Students stay in one classroom and the teacher must move. I think the main reason for this is to help students learn their pledge as a member of society. We have special class contests, for example, a chorus contest or an athletic day during the school year. A second reason for this is it may be easier to control. During class time, there's no one in the corridors--they are very quiet. Another reason is to help students build friendships. Middle school, particularly, is viewed as a critical period of time of intellectual, physical and emotional development. This is the time when students build their lifelong friendships.*

Sajin went on to share her impressions of middle school science teaching in South Korea:

> *We have just one textbook for science, and it's divided by subjects. The textbook is much smaller, does not contain as much information, as ones I've seen in the U.S. We separate by teachers--one teacher taught Biology sections and I taught Chemistry and Physics sections. We emphasize big ideas and scientific thinking. For example, we place a great deal of emphasis on using mathematics in science--in 7th grade they must be able to calculate Newton's laws and in 8th grade memorize the periodic table and understand basic chemical formulas. Students don't have many opportunities to do hands-on science. We have a*

separate lab room, and maybe two or three times a semester students can have a lab experience. The problem is the physical environment and lack of space and equipment. But I always tried to bring in some demonstrations.

In general, Sajin felt that both the Japanese and U.S. systems of education have served as models for South Korean schooling. However, she viewed teachers' general lack of pedagogy as a continual source of frustration. She reflected:

The government publishes a national curriculum guide that mentions ideas such as inquiry and STS [science, technology, and society], but it's just not done in the schools. Maybe it's because the science education methods classes are focused almost entirely on content and concepts. Science at both the secondary and university level is not taught in a way that relates to students' everyday lives.

Like our other interviewees, Sajin considered herself to be scientifically literate. She explained:

I'm scientifically literate because I'm comfortable using both science and technology. For example, take a person who needs to know more about global warming. If that person wants to learn more and knows where to go for more information--that's a big part of scientific literacy. As I get older I forget some science information, but I know where to go to locate that information. Also wanting to read current articles--staying current--is an important aspect of scientific literacy.

Sajin went on to describe other aspects of scientific literacy that she felt were visible in her husband:

When I ask him about a confusing concept in science, he explains very precisely and clearly in terms that I can understand. He has a very specific knowledge in physics but also a very general understanding of different sciences. And he's very excited about science.

Sajin acknowledged that in South Korea computer literacy is currently thought to be even more important than scientific literacy. She emphasized, however, that both were considered "basic to survival and getting a job, which pushes people to work hard":

I cannot overemphasize the extreme competitiveness of our society, which makes it very different from the U.S. If you want to be a teacher in Korea, it's very difficult to get a job.

When we asked Sajin about the meaning of the "science for all" slogan that is often associated with scientific literacy, she was passionate in her response:

> *When I first heard "science for all" I immediately thought it is very utopian, not realistic and even ridiculous! But after looking at the Standards [U.S.* National Science Education Standards *(National Research Council, 1996)], I began to wonder if this could be a realistic goal. I'm still struggling with the concept. When I discussed this with my husband he said 'why push everybody to be scientists?' Actually, in Korea, at one level, science educators want all students to become scientifically literate, but I think most people believe that it depends on individual interest and ability.*

When asked to suggest factors that might facilitate or constrain the development of scientific literacy, Sajin cited the school and the quality of the teachers as the most important factors in fostering scientific literacy for South Korean students. She went on to explain several other significant differences between schooling in South Korea and the U.S.:

> *In Korea the time spent in school is much longer than in the U.S. When I was in high school, it was not unusual to arrive at school at 7:30 a.m. and not leave until 10:30 p.m. The only time I was at home was to sleep. In terms of scientific literacy, another important factor is having a key "supporter." Sometimes it's having strong family support, but when that is not available, teachers must take on that role.*

We were surprised to learn that a separate gendered school system exists at the middle and secondary school levels in South Korea. Sajin explained:

> *We have boy classes and girl classes and the two are not mixed except at the elementary grades. It is like this because of cultural tradition, although it is slowly changing now. I remember student teaching. At that time I wanted to try a hands on lab and I prepared lots of materials. But the class was a "boy class" and they were all talking. I desperately struggled with control, not just the content. And that's probably another reason why I'm interested in pedagogy today.*

Sajin elaborated further on the structure of student teaching field experiences in South Korea, saying:

> *We go to "attached" schools for student teaching during our senior year. All students go at the same time and there are five or six of us with one teacher in the same class. First we went to the elementary for one week and maybe during this time I taught just two periods. Then we went to*

> *the middle or secondary schools. Each student teacher is responsible for planning and teaching several model lessons while everyone else observes, takes notes and provides feedback. I think our student teaching system is similar to what's found in Japan.*

As Sajin speculated on her future role as a science teacher educator in South Korea, she described some of the changes she would like to see:

> *My experience has been that science has always been taught separately in isolated pieces. I recently have observed some block scheduling. I would like to see more systemic structures such as this used in Korean schools. In the methods classroom, I want to emphasize scientific process and scientific ways of knowing, not just science content. I want to emphasize the importance of interpreting and using information and the development of an appreciation for science. And I would like to include PCK [pedagogical content knowledge]--because I believe it's important in terms of both doing science and learning to teach science.*

Hideo Ikeda: Japan

With the creation of a new national science curriculum in Japan, *Model for Japanese Education in the Perspective of the 21st Century* (Central Council for Education, 1996), Japanese educators and leaders are redefining science education. They believe that a critical need exists to shift science education from an emphasis on memorization and conformity to a focus on individuality, inquiry, and risk-taking:

> There is a need to change the style of education as we know it, and move away from a form of education which tends to favor the one-way inculcation of knowledge to a form which aims to cultivate in children the power to learn and think for themselves. (Central Council for Education, 1996, p. 134)

Considering that Japan is in the midst of a reform era whose goal is, among other things, achieving scientific literacy, we sought the perspectives and comments of a science educator, Hideo Ikeda, of Hiroshima University. Hiroshima University has a reputation of excellence in science education and is a leading teacher preparation institution in Japan dedicated to the professional development of practicing and prospective teachers.

Hideo Ikeda is currently a Professor of Science Education on the Faculty of Education at Hiroshima University. He was born at Yamaguchi Prefecture, approximately 150 miles west of Hiroshima, but moved to Hiroshima as a young child of six. Science has always been of interest to Hideo, even in his early years of schooling:

When I was a primary school pupil, I was interested in science and math. I can remember some hands-on activities in elementary school classes

One activity in particular reminded Hideo of changes that need to be made in today's schooling; it emphasized children asking their own questions, searching for their own answers:

When I was in 3rd grade in elementary school, we made water pistols using hollow bamboo stems. At first the water did not come out from my pistol because of water leaking. I had found out the reason why the water leaked: the inner diameter of the bamboo cylinder was not appropriate. The bottom diameter was wider than the top. Then, during the next class in the same week, I had another chance to replace it with a new bamboo and succeeded in stopping the water leaking. Now in Japanese elementary schools, teachers supply their kids with plastic air pistols. Using this kit, almost all the kids might succeed without leaking, but I think there also might be no chance to learn from a failure. Which is better?

As Hideo illustrated in subsequent remarks, science has always been a strong focus of his education:

In high school I was interested in biology, then I prepared to take an entrance exam for the Faculty of Education at Hiroshima University. I was very much interested in microscopic observation in high school. After finishing graduate school, I got a job as a research associate for the Faculty of Science. At first I was interested in biology research, but later I was much more interested in secondary school education. Then I moved to junior and senior high school. In the high school, I engaged in student teaching. Finally, I moved to university teaching.

Reflecting on a family history with strong ties to education, Hideo commented:

My father was a university professor and my mother was a high school teacher, I think it was natural that I was thinking to become a teacher.

Hideo noted that his strong educational background has prepared him to be scientifically literate:

I consider myself scientifically literate from the viewpoint of pragmatic thinking, an idea which was introduced from Europe and the United States into Japan more than 100 years before. I have been trained under this way of thinking from elementary school to graduate school.

For Hideo, scientific literacy involves a way of thinking and is closely connected to culture. Hideo discussed the connection he perceives between culture and science education in Japan:

> *I think the word science has almost the same meaning all over the world, but especially in under-developing countries it seems to comprise much more dreams, good hopes and goodness until now, and like developing time 1945-1960 in Japan when I was educated. I think the meaning of goodness in science has been changed or decreased through the Cold War and Japanese development. On the other hand, 'science education' has a little bit different meaning in every country, especially between western countries (Europe and U.S.) and East Asian countries (China, Japan and South Korea) because it may have been influenced not only by the science content, but also history, culture, tradition, or policy of each country. For example I think, you had reclaimed your broad land from the East Coast to the West with American pioneer spirit. I think you Americans have conquested nature, settled the undeveloped land. We Japanese have been living in such narrow land for more than 2000 years. We could not live without harmonizing with our small nature and environment. I have been investigating your* Science Education Standards *[National Research Council, 1996] and* Benchmarks for Scientific Literacy *[American Association for the Advancement of Science, 1993] and the* Japanese Course of Study *[Central Council for Education, 1989] stipulated by the Japanese government. In our Course of Study, I have found out one important goal in elementary school science education is 'love for nature.' I think this goal is one of the most distinctive features in Japanese science education compared to the U.S. and Europe. This notion had appeared from 1891 in the goals of Japanese elementary school science education and it has been lasting until now. In Japanese elementary schools every kid may be required in taking care of their own plants or animals at school, and are required to clean up their classroom, schoolyard, school garden, private rooms, etc. We have been regarding this as a student's duty. Usually we have shared school hours for such special duty from elementary to high school. Your schools have been maintained very clean and well equipped, but it has been done by laboring people.*

For Hideo, scientific literacy is also characterized by skills. He remarked that a scientifically literate person "can solve her/his problems which they may encounter everyday with scientific way." Additionally, he stated that motivation, interest in science, and even necessity for one's career may affect one's state or level of scientific literacy.

Interestingly, when Hideo reflected on the meaning of scientific literacy, the first images that came to his mind were ones that contrasted the "American dream" with the current social climate in which he feels Japanese children are immersed:

At first I imagine "American dream." American kids seem to have many more dreams than Japanese. Now the Japanese students, especially in junior high schools, are suffering some problems: suicide, violence, spiritless, dislike for science. I think they are tired in study, hopeless in their future. Before 1980, we believed that scientific literacy brought our good futures, but the belief had vanished away by the 1990's. This is the most crucial problem for us.

As Hideo's comments indicated, efforts to achieve scientific literacy may not always have positive outcomes on society. Linking the notion of "science for all" to scientific literacy, he continued:

I would like to think "scientific life for all." We have succeeded in cramming science into kids, especially in Japan. Unfortunately, their scientific knowledge is only a tool for making high scores on entrance exams. Their knowledge cannot apply to everyday lives.

Despite what Hideo perceived as a "misuse" with respect to the goals of scientific literacy in Japan, he remained optimistic about the future. He explained his role as a science educator in facilitating goals associated with scientific literacy:

It is a chance to change the situation. I think our students must be taught how to solve problems which they may encounter in their everyday lives. We teach students with precious equipment and exquisite procedures, but I would like to add some activities using kitchenware or junk in a trash box with simplified procedures. I call it "trash box biology." For example, when we need a thin-cut sample for microscopic observation, we usually use microtome, but we can make it from an empty lipstick. I am encouraging my students to find out their own problems from the kitchen.

Elizabeth McKinley: New Zealand

Elizabeth (Liz) McKinley was born in a small rural town named Carterton in Wairarap, Aotearoa, New Zealand. As a Maori, her tribal affiliation is Ngati Kahungunu and Ngai Tahu. Her professional career began when she received her Bachelor of Science degree in Chemistry from Otago University. She taught secondary science for 12 years--6 years at a co-ed school and 6 years at a girls school where she was also Head of Science. She then became a preservice educator at the University of Waikato in Hamilton, Aotearoa, New Zealand, where she was responsible for teaching primary and secondary science and science education programs. For the past three years, Liz has been a Lecturer in science education at the Centre for Science, Mathematics and Technology Education Research at the University of Waikato. She teaches master's courses in research methods, science education, curriculum policy and development, and supervises masters and doctoral thesis students. Her current research interests include the influence of culture and language in science and science education, feminism and science education, and curriculum policy and development.

Liz describes herself professionally as a "Maori woman with a strong interest in feminism and science, and culture and science/education." For Liz, her professional involvements have had tremendous personal implications:

> *I see myself as part of the science education community I critique, but also I see myself outside this community as well as part of the Maori community who can (and do) critique the science/science education community. I am both, and this is one example of the external contradictions I live with, and live out, everyday.*

Liz's sense of contradiction has led her to challenge deep assumptions associated with science and science education.

In her definition of scientific literacy, Liz calls for attention to the cultural, political. and historical contexts of life in New Zealand:

> *Scientific literacy culturally begs the question, 'Who is scientifically illiterate?' The "western" scientist who has no understanding of Maori views of nature? The "non-westerner" who doesn't know of the 'green revolution'/Gaia theory?...Our problem is scientific literacy of the Pakeha (white) population with respect to the Maori views of nature.*

The political relationship between these two groups has resulted in the subjugation of Maori beliefs, as Liz explained:

> *In Aotearoa, New Zealand we are a 'developed' nation with an indigenous population that is recognized as such. However, we have a history of the Pakeha (white) population being dominant in almost every sphere of social and political activity. Our issue is to get recognition in cultural practices in science (i.e., the scientific literacy problem). Whilst all Maori have to learn 'Pakeha ways' (by the very nature of their dominance in our country), the reverse does not have to happen.*

Liz followed with a vignette that underscores this dilemma:

The kiore, a species of rat, is a cultural taonga (prized possession) for both genealogical and historical reasons. In some Maori accounts, the kiore can be found in the creation traditions of our people and is closely related to humans through genealogy. About 1000 years ago, the kiore was brought to Aotearoa in ancestral canoes. To some Maori, the kiore rat was an historically important food source. In a land that has no indigenous terrestrial mammalian fauna, the value of this resource extended well beyond nutritional necessity and assumed importance in virtually every other aspect of Maori life. Kiore were used as currency in the acquisition of land, and in the naming of people and places. They were also sung about in waiata (songs) and haka (a dance of challenge), and knowledge of their habits was incorporated into whakaatauaki (proverbs) as well as depicted in carvings. With the introduction of the larger European associated rats, along with stoats, ferrets, weasels, and cats, the kiore gradually disappeared from the two main islands to some offshore islands. Unfortunately for the kiore, many of these islands also contain remnant populations of the endangered indigenous plants and animals.

Research into the impact of rodents on the flora and fauna on many of these island reserves implicated the kiore with the other rat species in several extinctions and the decline of many other animal and plant species. Based on this evidence, the Department of Conservation embarked on an eradication program of all rats including the kiore. By the time Maori groups got to hear about it in early 1992, six island populations had been eradicated. It took two years and the intervention of New Zealand's Commissioner for the Environment to respond to Maori calls for a management plan for the kiore instead of an eradication plan.

This situation was particularly seen as an affront to Ngatiwai, the tribe who are the kaitiaki (guardians) with mana whenua (traditional authority and responsibility for the islands involved). Ngatiwai are wanting to do research with respect to the kiore and their lands, but are finding it frustrating when dealing with the New Zealand science system. One of the issues Maori still face is that Maori applications for the Public Good Science Fund (currently the biggest science research funder in New Zealand) are being turned down. Maori questions relating to science, their proposed resolutions and the localized nature of the research proposed are clearly being seen as less valid by assessors with different ideas about best research composition.

When we asked Liz about the educational slogan of "science for all" and the implications this might have for dealing with diverse views of science and science education, she responded:

> *I'm a bit skeptical about the notion "science for all," particularly how it has been taken up in science education literature and research. For my own situation, I treat it as a "western" science education myth! I see it as a way the "west" can further promulgate their ideas, values, etc. to non-western peoples and/or countries. The phrase has a lovely sound to it, appealing to a 'notion of universal humanity' (totally consistent with a liberal humanist position), which of course appeals to a wide number of people. This is consistent with a straight multiculturalist position in science education but is not necessarily supportive of more particularized positions, such as the biculturalism advocated in Aotearoa, New Zealand, and the structural barriers such groups face. The term biculturalism in New Zealand has to do with partnership between Maori, the indigenous group of this land, and the European settlers that came in 1840. There was a treaty signed at this point to allow European (British) settlers into the country. Biculturalism is a discourse that is about constructing our identity as a country, one which reflects this dual heritage, in all its social institutions and structures. This discourse is separate from the multicultural one, which includes the two groups above plus other immigrant groups that have come to the country in different circumstances and other times. This may be the same for other indigenous groups (e.g., Native Americans, Latinos) who are fighting to have their own cultural norms recognized in their own countries...No one that I have seen using the phrase "science for all" has really asked or theorized about the hard questions like: Whose science? Whose knowledge? Who gets to do science?*

When we asked Liz to comment on the role of schools in promoting scientific literacy in New Zealand, she described both the nature of science education and its status within the community:

> *Everyone in Aotearoa, New Zealand has access to schooling and we have compulsory schooling until 17 years of age. I think schools do have a role, and quite an important one, in promoting scientific literacy. At the same time, there should be more credence given to informal structures and making use of the 'scientific' community as a whole--not just trained scientists! There is a place for formal science training, but I think it should be mostly left to tertiary institutions, and maybe senior secondary schools. I think primary and secondary schools (particularly the 11-16 age group) should broaden their programs and make them more relevant*

(culturally and contextually) with more local input. However, there is only so much a school can do under the current educational structures.

I think that the entire community has an important role to play--from the media, to scientists in the local research institute, to the local elder that has fished all his life. I think our first problem is that the community has to value this knowledge in a way that is different from the high status science carries currently. Science, as we currently know it, has a very authoritarian 'voice' in our communities. This has to change. Some of it has to do with the public 'image' of science, and some about learning how science is culturally, historically, socially, and economically constructed.

When we asked Liz to describe how one might measure or evaluate a person's scientific literacy, her response focused more on what a scientifically literate person might be expected to know:

If they can communicate with their local community--that would be a start...For us in Aotearoa, New Zealand, environmental/ecological knowledge is very important--it is important economically for our image to be "green." It has been, and is still, important for us historically as England's farm for food production. Also, as a land that has one of the largest flightless bird populations in the world. And, it is important culturally, as the role of land in Maori views of nature is paramount. These are but a few reasons.

Liz began to consider the possibilities of scientific literacy viewed from a localized to global perspective:

I think that some knowledge needs to be even more particularized--a community level where knowledge of a particular beach, forest, lake whatever and how locals have historically looked after it, abused it, etc. And then, of course, there is an international level--knowledge put in a wider picture. I'm not sure how you do this, any of these really. Do we want to? For what purpose will we measure? As soon as we bring in measuring and/or judgments, we're back to having to 'use' universalist notions of what becomes science. Can we agree on what are world-wide issues (this of course will change over time) and take our cues from there?

Liz's concluding questions and examples contribute an important basis for science educators to reflect on the assumptions of achieving scientific literacy--what will be gained and what will be lost--and to whose benefits or detriment?

Lesley Parker: Australia

Lesley Parker is the Senior Deputy Vice-Chancellor at Curtin University of Technology in Perth, Australia. Previously, Lesley was a professor of higher education in Curtin University's Science and Mathematics Education Centre where she served as Associate Director of the National Key Centre for School Science and Mathematics (Especially for Women). She has spent twelve years in postgraduate science education and continues to supervise the research of doctoral candidates. Throughout her career, Lesley has been a leader in forging equity-related policies and research in science and mathematics education. Accordingly, Lesley's views of scientific literacy expressed in this paper reflect her long-term involvement in these areas of educational policy and academic research.

Lesley's experiences in negotiating equity issues (most specifically, gender-equity) in educational policy initiatives and in science education research have deeply impacted her thinking about the purposes of science and science education. "My thinking on the role of schools in developing scientific literacy is no doubt influenced by the documentation we have produced and the debates we have had."

For the most part, Lesley perceived that her views were not reflective of "mainstream" thinking about science and science education. The insufficient manner in which traditional theory and research have responded to equity issues overall have compelled Lesley to explore feminist literature and research methodologies:

> *My perspectives are derived from the work of feminist scholars such as Nancy Tuana (1989) who have challenged traditional definitions of science and scientific rigor in their attempts to 'humanize' the curriculum. In my work, I've developed a sociohistorical analysis of a number of examples of such attempts--specifically those associated with the development of "science for all" curricula in the 1960s-1970s, and gender-inclusive curricula in the 1980s. A major dilemma with these curricula is that they tend to be perceived as 'not science' or as a lower status science--apparently because of the alternative 'ways of knowing' and the alternative concepts of science embedded in them. The concepts presented by Mary Belenkey and her colleagues [Belenkey, Clinchy, Goldberger, and Tarule, 1986] regarding 'women's ways of knowing' have been particularly helpful toward looking at situations of minority cultures in Western societies and the accommodation of minority cultures in Western science.*

Overall, Lesley's interpretations of 'scientific literacy' reflect these feminist perspectives:

> *When I think about my overall impression of the phrase 'scientific literacy' it brings to mind lots of meaningless rhetoric from those who hold a very narrow definition of science and of scientific literacy. In many cases, words that might be used when people talk about science such as, "evidence," "open-minded," "knowledge," and "understanding," I see a completely different interpretation of the meanings of those words which can be equated with domination and control.*

As she responded, Lesley emphasized the need to critically examine the perspectives and purposes served under the rhetoric of scientific literacy. In her critique, she began by characterizing scientific literacy as involving four "attitudes of mind" with the first featuring a "curiosity about the physical and biological world and a wish to understand this world." A second attitude concerned the notion of "evidence":

> *Conclusions, inferences, etc. need to be backed up by evidence. But, at the same time, an understanding that the evidence, itself, must be open to question in terms of the paradigm and assumptions underpinning the way in which it was gathered and the questions that were asked as part of the gathering.*

Lesley described a third attitude that highlights the multiple perspectives people bring to interpreting their world:

> *Thirdly, it's a willingness to remain open-minded, in the sense that there is no one way to explain the world, and that even apparently logical and readily accepted explanations represent only the culmination of the state of knowledge at a particular time and in a particular cultural context.*

Furthermore, she added, "such explanations must be seen as constructed by men (and sometimes women) --human beings who themselves are constructed by their cultures (academic and social)." The fourth attitude she highlighted was: "A preparedness to take a responsible role in relation to the use of science and its applications in everyday life."

Lesley also shared that, for her, scientific literacy also involves a communicative capacity: "Scientific literacy involves the classical language-related skills of reading, writing, speaking, listening and viewing, and it involves the exercising of these skills in a critical way." Consistent with the critical perspective she has long practiced in her work, Lesley called attention to her reference to "language" to add that:

> *I am using 'language' in the generic sense--that is, language is not equated with English. It is not equated with the use of formalized modes of expression seen by some as a necessary characteristic of science, and*

> *it is not equated with the quantitative and with the substitution of symbols for proper nouns.*

Lesley's explanations regarding her interpretation of scientific literacy served as an important prelude to the subsequent discussion about the role of scientific literacy in education. She explained:

> *Please note that I do not think one has to "do" science in order to demonstrate scientific literacy. I like to think that I personally am scientifically literate in that I have all of the above (i.e., four attitudes of mind and communicative capacity) at least to some degree. I see my own scientific literacy as demonstrated through a kind of critical discourse analysis, involving critique of the scientific content of a piece of discourse (or an event, etc.) and critique of its underlying premises and paradigms. Almost anyone CAN be scientifically literate. They do not need to have studied "science" as such, and they do not need to be practicing science. In fact, my definition would probably exclude a number of practicing scientists.*

Her comments regarding the notion of "science for all" are indicative of why and how Lesley sees "critical analysis" as having a central role in scientific literacy:

> *When I hear the phrase "science for all" it means, to me, a view that "science," whatever that might mean, must be part of the education of all citizens--not just of an elite group who are destined to become practicing scientists. And that curricula, both formal and informal and involving education from early childhood through to adult learning experiences, need to be devised to enable this to happen.*

Lesley sees a caveat, however, in the notion of "science for all," in that:

> *It could lead to a dilution of the curriculum required for progress to scientific careers, or to a premature specialization of those students aspiring to such careers. It can also lead to a situation where "real" students (usually male, white and middle-class) are seen as doing "real" science, while "others" (i.e., females, etc.) do "science for all."*

Lilia Reyes Herrera: Colombia

Lilia Reyes Herrera is a science educator and professor of biology and education at the Universidad Pedagogica Nacional in Bogota, Colombia. She has taught for nearly twenty years. Lilia described her own beginnings in education:

I was born in a very small town. The name is Velex, Santander. The temperature is warm and there are a lot of guava plantations nearby. Most of the people are very honest, appreciate and follow Catholic values, and are very hard workers. I have four sisters and two brothers. I am the oldest. One of the roles of the oldest girls in the family is to help with the younger children and the parents. All my brothers and sisters went to the university and graduated. This situation is unusual in our context, but it was not impossible, thanks to our parents, especially my mom, an eager woman who knew that education is the best treasure you can give your sons and daughters.

We asked Lilia how and why she became involved in science education. She responded:

I do not know exactly why, but I knew that I wanted to be a good teacher. After I finished high school, I passed the exams to enter into the most prestigious public university of the country to study Licenciatura in Biology and Chemistry. So I was certified to teach at the secondary level.

Lilia began her teaching career as a part time biology professor and also worked in the Adult Education Center:

While teaching at the university, I continued studying for the master's degree in Agricultural Sciences and I also went to Boston College for a Certificate of Advanced Educational Studies. I also won a scholarship from the Netherlands government to study plant breeding for four months in Wageningen, Holland.

In her early days at the university, Lilia recalled talking with colleagues about illiteracy among the non-academic workers at the university. Lilia and her colleagues wondered, "How can we help to change the situation of these workers? We realized there were people right there on the campus every day who could not read or write, such as the gardener. Some of them had to stamp their fingerprint when receiving their paycheck since they were not able to write their signatures." Lilia and her colleagues became enthusiastic about finding solutions. They decided to begin a free-cost literacy program.

About twenty adult workers participated as students during the time they were not on duty. A number of custodial workers, people who work preparing students' food at the cafeteria, electricians, and maintenance workers began attending on a regular basis. There were people from all ages, from 18 years old to more than 50 years old. Many have learned to read through this program and continued studying to obtain the equivalent of the elementary schooling. More students came, and some of them continued to attend the Adult Education Center, which was attended mainly by student teachers in all disciplinary areas.

To date, some have finished their high school studies, and at least 2 have received bachelor's degrees from the university as a result of their participation in the program.

After completing her master's studies, Lilia's interest in becoming a good teacher continued to grow. She also had a desire to impact science education in a larger way. She obtained a Ph.D. in Science Education from Florida State University and is now involved with science education at her university. When asked to describe what science classrooms are like in Bogota, Lilia responded:

I think that there are not "typical" classrooms now. All are very different. There is a big change tendency in most of the schools that I am attending and doing research on teacher's beliefs about the nature of science, the nature of learning, and the nature of teaching. The new Constitution and the new Law of Education [Colombian Constitution Law 115, 1994] gave educational institutions an opportunity to construct their own pedagogical projects. [The constitution of 1991 is the official document that contains a system of laws and principles that govern the Colombian State. The Law of Education, properly called Ley 115, is a set of rules that regulates education as a personal process of formation, personal, cultural of human person, dignity, and social based on the integral conception duties and rights.] So more teachers are working hard to contribute to the scientific preparation of Colombian human capital. I can say that maybe the typical aspect could be the change and the evolution from traditional ways of thinking and acting about education to new alternatives.

As Lilia reflected on the importance of scientific literacy to Colombians, she indicated that this is a very important educational goal in her country. Lilia emphasized that scientific literacy has many possibilities for enhancing how people learn and live in society. She described science literacy as: "An important way to understand the world around us. It enables people to get more meaningful and deep understandings of phenomena." Further, she stated that scientific literacy is essential to the intellectual growth of persons:

Scientific literacy is part of human development. It mostly involves intellectual and material well-being. A scientifically literate person has the foundations to understand phenomena and people in a different way when they talk about science. Scientifically literate people can make reasonable explanations of events making predictions about problems and use coherent methodologies to solve them.

As Lilia continued to describe images she associates with scientific literacy, she emphasized its potential for enhancing personal and social understanding:

> *When I think of scientific literacy, I think of progress, understandings, advantages, sharing meanings, gaining access to new communities of thought and action, systematization and clarification of thoughts. People who have previous experience in scientific thought and people who have fundamental skills to be able to understand and interpret scientific issues. People who can participate with others in meaningful conversations about the world. People who have the abilities to understand scientific events. People acting together to improve life and well-being.*

She contemplated:

> *Can we say that scientific literacy is innate or that it is the product of schooling? I'm thinking here about the situation of many people, especially in the rural areas, who without going to school are able to state problems, find ways to solve them, and find solutions that work in their context.*

Lilia sees scientific literacy as an important resource that can expand the ways people interpret and talk about their world and how they want to live in it.

Reflecting on the notion of "science for all," Lilia expressed her belief that scientific literacy is essential to democratic society:

> *"Science for all" means possibilities of democratic practices. Sharing common ways to solve the problems and to find solutions. Sharing possibilities. Contributing to make a better world for more people. Equal opportunities to access science, to belong to a new culture.*

Lilia sees disadvantages associated with the slogan "science for all" and stated that: "A very negative aspect results when scientific progress and results are [illegible] in the exercise of human values. It is not wise to consider science independent from other possibilities of integral human development." While she did not underestimate the important contributions that other ways of knowing might contribute in society, Lilia described science as a more expansive and well-informed way of thinking:

> *I think that scientifically literate people are more tolerant, less dogmatic, and possess a lot of good reasons for us to understand events and to take a position. Scientifically literate people are more open to finding new solutions to new problems and with new methodologies. They can also identify more relevant problems and give more alternative solutions.*

They can document and give a lot of evidence to support the assertions they make.

Lilia is a strong advocate for promoting scientific literacy through whatever public means are possible. As a science educator, she described her own role in promoting scientific literacy through:

[Being a] facilitator, motivator, guide, a knowledgeable person who finds and practices different opportunities for students to think about science in a non-threatening way. I provide plenty of opportunities for students to have scientific skills and knowledge to deal successfully with problems and issues in everyday life. Educational institutions should be places that allow learning and practice towards the development of scientific thought. They should be gathering places to motivate and facilitate students' scientific thought development.

She believes, however, that a student should not be "forced against her or his will to study science courses." She emphasized that other places besides schools can serve an important role in promoting scientific literacy:

People should have rich opportunities to talk about science; to be valid peers to discuss events...Institutions such as museums, botanical gardens, parks, can make a big difference in the way people think and act, if they can have scientific explanations for what is happening in that place.

Lilia's responses resonated with those expressed by Jeff Thomas (1997) when he said:

> Individuals are not passive recipients of raw science--neither is science seen as a superior form of natural understanding. Knowledge of science becomes situated within a broad web of complex social and emotional factors.... A scientifically literate public is perhaps one where science attracts an emotional mix of dispute, ambivalence, anxiety and appreciation in ways no different from other human pursuits. (p. 172)

Understanding complex issues that challenge the promotion of scientific literacy, and literacy in general, remain central to Lilia's ongoing research and teaching efforts. Currently, her research is focused on developing university and school partnerships. Lilia noted that educational policies have recently been developed that call for a better understanding of issues that challenge classroom teaching and learning. According to the document Lineamientos Generales de los Procesos Curriculares (General Guidelines for Curricular Process) the Colombian educational system is exploring a deep transformation in education: "El hecho de que hayamos sido formados con otra optica y en otras circunstancias no nos exime de la responsabilidad de construir juntos la escuela que las circunstancias historicas requieren" (Ministerio de Educacion Nacional, 1994, pp. 20-24). Lilia translated this as: "The fact that we were formed with a different vision and in different circumstances does not free us from the responsibility to construct together the school that the historical circumstances require." Lilia proposed that:

> *A significant alternative to consider, specifically in science education, would be the identification of social barriers that represent important obstacles to learning for teachers and students. It might be possible that by identifying cultural myths and consciously reflecting on them, Colombian teachers would have a different and powerful vision of education.*

IMPLICATIONS FOR SCIENCE TEACHER EDUCATION

There is a sense of tension as endeavors to globally standardize and evaluate science education (e.g., the *Third International Mathematics and Science Study*, [Beaton et al., 1996]) press forward. For some, the banners of "scientific literacy" and "science for all" are seen as essential to a democratic way of life; others see these slogans as ideological threats. The issues embedded in our participants' responses hint at the complexities facing science teacher educators internationally as we attempt to move beyond arcane notions of science and science teaching to facilitate meaningful science learning for all students.

One of the perspectives highlighted in this study is that what is considered meaningful to teach and learn is relative to what is seen as significantly important in people's everyday lives. Dialogue promising enhanced quality of life brought about through the development or attainment of scientific literacy reflects a deeper assumption that the world is deficient and undifferentiated (Levinson & Thomas, 1997). We can see in several of the narratives how cultural, historical, moral, and political issues are at stake. In Lilia's world there is a strong desire to maintain democracy; for Liz there is the struggle to preserve Maori culture and local resources; for Hideo there is the struggle to maintain an age-old historical relationship to the land. Scientific literacy, when viewed as a means for individuals and groups to construct knowledge about their world, is constitutive of the social, cultural, and political commitments of local communities. Careful consideration must be given to the rhetoric of globalized scientific literacy and the power of local communities to construct knowledge in ways they see as essential to their future well-being.

In science teacher education, there must be an awareness that there are, indeed, alternative needs and ways to think about situations in everyday life. The pedagogies of Science Technology and Society (Solomon & Aikenhead, 1994) and "female-friendly science" (Rosser, 1997) challenge curricular assumptions about the role science plays in society and who participates in doing science. However, more work is needed to analyze monolithic references to "science" and the social and political implications of promoting scientific literacy. Opportunities are needed to consider how "science" relates to other ways of knowing, and the purposes different knowledges serve toward making sense of life experiences. This raises important questions about what teachers of science see in the futures of their students and what knowledges are believed critical to their future well-being. This can be problematic, given that the life experiences of teachers do not necessarily reflect those of their students (Aikenhead, 1996), much less those who are not consistently involved in school (Barton, 1998). This has several implications for teacher preparation. First, greater efforts must be made to recruit future teachers who have substantive experiences learning in culturally diverse settings. This might involve informal or formal teaching experiences in rural and urban settings, high and low-income settings, and in settings that contrast with their own ideological origins.

Additionally, science teacher educators need to consider the implications of moving beyond hegemonic representations of scientific literacy related to their own teaching practices. Science teacher educators have introduced a variety of "critical tools for reflection" (Nichols, Tippins and Wieseman, 1997), intended to assist prospective and practicing teachers in critically reflecting on their assumptions of what constitutes science knowledge and teachers' roles in science teaching; the narratives in this chapter are an example of such "tools." They reflect the writers' underlying cognition, express their pedagogical beliefs, and represent ways of perceiving and organizing reality. Others have explored pedagogies that challenge teachers to examine how sciences are personally and socially constructed and are therefore shaped by the cultural and political interests of those involved (e.g., physics--Roychoudhury, Tippins and Nichols, 1995; geoscience--Mayberry & Rees, 1997; life science--Rosser, 1987). Some have ventured into interdisciplinary work to develop "hybrid pedagogies" (Hildebrand, 1998) to enable teachers and students to explicitly examine hegemonies embedded in representing the natural sciences as isolated disciplines. Deborah Pomeroy (1996) presented an especially useful overview of nine agendas for addressing science education teaching, policy, and research respective of cultural diversity. Increasingly, science educators are developing pedagogies that shift away from the assumption that science is a term having universal meaning and application.

Ultimately, there is a need for deeply insightful conversations about what and whose purposes are served under the guise of educating for "scientific literacy." Currently, such conversations privilege those with access to communication technologies, texts, journals, and conferences. How can opportunities be created to learn from the views of those who have historically not been heard in such discussions? Opportunities need to be created to extend resources that help diverse communities think about what it means to be scientifically literate. Many countries and states have recently engaged in the production of curriculum standards; this perhaps presents an opportune time to discuss within and across communities what these documents have to say about the purposes and practices of scientific literacy.

REFERENCES

Adams, H. (1918). *The education of Henry Adams.* Boston: Houghton Mifflin.

Aikenhead, G. (1996). Science education: Border crossing into the subculture of science. *Studies in Science Education, 27,* 1-52.

American Association for the Advancement of Science. (1993). *Benchmarks for Scientific literacy.* New York: Oxford University Press.

Barton, A. C. (1998). Teaching science with homeless children: Pedagogy, representation, and identity. *Journal of Research in Science Teaching, 35,* 379-394.

Beaton, A. E., Martin, M. O., Mullis, I., Gonzalez, E. J., Smith, T. A., & Kelley, D. L. (1996). *Science achievement in the middle school years: IEA's third international mathematics and science study (TIMSS).* Boston, MA: Center for the Study of Testing, Evaluation, and Educational Policy.

Belenkey, M. F., Clinchy, B., Goldberger, N. R., & Tarule, J. M. (1986). *Women's ways of knowing.* New York: Basic Books.

Boyer, E. L. (1990). *Scholarship reconsidered: Priorities of the professoriate.* Princeton, N. J.: Carnegie Foundation for the Advancement of Teaching.

Central Council for Education. (1989). *Course of study for elementary school.* Tokyo, Japan: Ministry of Education, Science, Sports and Culture.

Central Council for Education. (1996). *The model for Japanese education in the perspective of the 21st century.* Tokyo, Japan: Ministry of Education, Science, Sports and Culture.

Clark, B. R. (1987). *The academic life. Small worlds, different worlds.* Princeton, N. J.: Carnegie Foundation for the Advancement of Teaching.

Colombian Constitution Law 115. (1994). *General law of education.* Bogota, Colombia. (In Spanish)

Ducharme, E. R. (1993). *The lives of teacher educators.* New York: Teachers College Press.

Hildebrand, G. (1998). Disrupting hegemonic writing practices in school science: Contesting the right way. *Journal of Research in Science Teaching, 35,* 345-362.

Kagan, D., & Tippins, D. (1995). Perceptions of education in the United States by nonnative professors of education, *The Educational Forum, 59,* 140-153.

Levinson, R., & Thomas, J. (1997). Science, people and schools: An intrinsic conflict? In R. Levinson & J. Thomas (Eds.), *Science today: Problem or crisis?* (pp. 1-5). New York: Routledge.

Mayberry, M., & Rees, M. (1997). Feminist pedagogy, interdisciplinary praxis, and science education. *NWSA Journal, 9,* 57-75.
Ministerio de Educacion Nacional. (1994). *Lineamientos generales de los procesos curriculares.* Bogota, Colombia: Author.
Mishler, E.G. (1986). *Research interviewing.* Cambridge: Harvard University Press.
National Research Council. (1996). *National Science Education Standards.* Washington, DC: National Academy Press.
Nichols, S., Tippins, D., & Wieseman, K. (1997). A toolkit for developing critically reflective science teachers. *Journal of Science Teacher Education, 8*, 77-106.
Pomeroy, D. (1994). Science education and cultural diversity: Mapping the field. *Studies in Science Education, 24,* 49-73.
Rosser, S. (1987). *Teaching science and health from a feminist perspective: A practical guide.* Elmsford, NY: Pergamon Press.
Rosser, S. (1997). *Re-engineering female friendly science.* New York: Teachers College Press.
Roychoudhury, A., Tippins, D., & Nichols, S. (1995). Gender-inclusive science teaching: A feminist-constructivist approach. *Journal of Research in Science Teaching, 32*, 897-924.
Shamos. M. (1995). *The myth of scientific literacy.* New Brunswick: Rutgers University Press.
Solomon, J., & Aikenhead, G. (Eds.). (1994). *STS education: International perspectives on reform.* New York: Teachers College Press.
Thomas, J. (1997). Informed ambivalence: Changing attitudes to the public understanding of science. In R. Levinson & J. Thomas (Eds.). *Science today: Problem or crisis?* (pp. 163-172). New York: Routledge.
Tippins, D., Oliver, S., Nichols, S., Kemp, A., Rascoe, B., Chun, S., & Li, H. (1998, January). *Scientific literacy: Exploring the metaphor.* Paper presented at the annual meeting of the Association for the Education of Teachers in Science, Minneapolis, MN.
Tuana, N. (Ed.). (1989). *Feminism and science.* Bloomington: Indiana University Press.
Wisniewski, R., & Ducharme, E. R. (1989). (Eds.). *The professors of teaching.* Albany: State University of New York Press.

NOTES ON CONTRIBUTORS

Sandra K. Abell is Professor of Science Education at Purdue University, USA, where she teaches preservice and practicing elementary teachers and graduate students in Science Education. Her research focuses on becoming a teacher of science, from the teacher preparation program, into the beginning years of teaching, and throughout a teacher's career. Much of her work involves collaborating with classroom teachers in sharing the work of teaching science.

Bah Amadou is Professor of Chemistry at the Universite de Conakry in Guinea, West Africa. He recently completed a master's degree in Science Education and is currently involved in teaching and science education reform initiatives at his university.

Ken Appleton teaches preservice and practicing primary and elementary teachers as well as graduate students in science and technology education at Central Queensland University, Rockhampton, Australia. His research interests include constructivist views of science learning, teacher professional development in science, and the development of science pedagogical content knowledge in elementary teachers.

Kate A. Baird is President of Designs on Learning, USA where she provides workshops and training to schools and businesses on designing and assessing learning processes.

Stuart C. Bevins is a science instructor and researcher for the Science Centre at Sheffield Hallam University, UK. His research areas are teacher preparation and the stages of professional development of teachers.

Ronald J. Bonnstetter is Director of Secondary Science Education at the University of Nebraska-Lincoln, USA. His research interests include effective teacher preparation, the scholarship of teaching, and the processes necessary for science education reform efforts to have lasting impact.

Lidia Borghi is Professor of Physics at the University of Pavia, Italy, where she graduated in 1962. She teaches undergraduate and postgraduate students of Physics and Mathematics who want to become teachers. Her research interest is Physics Education and, in particular, the use of new technologies in physics teaching.

Saouma BouJaoude is Associate Professor of Science Education at the American University of Beirut, Lebanon, where he teaches preservice and practicing middle and high school teachers and graduate students in science education. He is the Director of the Science and Math Education Center at the same university. His research interests focus on science teacher education, students' alternative conceptions, and chemistry education.

J. Mauro Briceno-Valero is Professor of Science at the Universidad de Andes in Merida, Venezuela.

Marilyn M. Brodie is a science instructor at Sheffield Hallam University and a researcher at the Science Centre in the UK. Her research continues into teacher preparation and the PRISM project.

Lynn A. Bryan is Assistant Professor of Science Education at the University of Georgia, USA. Her research interests include teacher beliefs, teacher knowledge, and international science education. She recently spent six weeks in Hiroshima, Japan learning and conducting science education research in Japanese elementary schools. She is also involved in a collaborative science education reform project with educators in Honduras.

Angela Calabrese Barton is Assistant Professor of Science Education at the Center for Science Education, University of Texas at Austin, USA. She directs USE-IT (Urban Science Education with Integrated Technologies), a National Science Foundation funded research and activist program designed to better understand the scientific literacies and needs of children and their families in poverty. She is author of *Feminist Science Education* (New York, Teachers College Press).

Pamela G. Christol is Educational Specialist with the National Aeronautics and Space Administration (NASA) educational program housed at Oklahoma State University, USA. She provides inservice training on NASA materials for K-12 educators. Her research centers around the use of technology in science education with an emphasis on the GLOBE Project in the US and UK.

Sajin Chun received her bachelor's and master's degrees in Earth Science Education from Seoul National University, South Korea. She is currently completing her doctoral studies in Science Education at the University of Georgia, USA. Her research interests include the nature of science, teacher beliefs, and Earth science education.

Anna De Ambrosis is Professor of Physics in the Physics Department and in the graduate School for Physics and Mathematics Teaching of the University of Pavia, Italy. She is involved in studies on science education in primary school, in physics education in high school, and in teacher preparation.

Hernan Garcia is a former professor of biology and currently a life science teacher at the Cochabamba Cooperative School in Cochabamba, Bolivia. He has been a leader in Bolivian educational reform for the past seven years, working with math and science teachers throughout Bolivia.

Ian S. Ginns teaches science education courses at the Queensland University of Technology, Brisbane, Australia, to preservice elementary teachers. He has a particular interest in the teaching of astronomy. His research interests include the study of attitudes and self-efficacy of preservice science teachers and beginning teachers. He has an interest in technology education and is involved in research with graduate students in this area.

Lilia Reyes Herrerra is Professor of Biology and Education at the Universidad Pedagogica Nacional in Bogota, Colombia. Her research interests include sociocultural dimensions of science teaching and learning and issues related to the development of scientific literacy.

Hideo Ikeda is Professor of Science Education on the Faculty of Education at Hiroshima University, Japan. He recently served as the Project Co-director for a 3-year science education research exchange project involving faculty from Hiroshima University and the University of Georgia. His research interests include biology education and science education reform.

Hafiz Muhammad Iqbal is Assistant Professor of Science Education at the Institute of Education and Research, University of the Punjab, Lahore, Pakistan. He has taught biology at the higher secondary level and botany at the graduation level, and has worked as Inservice Educator for science teachers. His research and writing focuses on preservice and inservice science teacher education.

Amy M. Jacks is a fifth grade teacher at Spring Run Elementary in Chesterfield County near Richmond, Virginia, USA. In her beginning years of teaching, she is implementing the experiences and knowledge gained during her teacher preparation program at Purdue University. Her interest in science education remains an emphasis in her development as an educator.

Janice Koch is Associate Professor of Science Education at Hofstra University on Long Island, New York, USA. She directs the Master of Arts Program in Elementary Education with a specialization in Mathematics, Science and Technology. She is author of *Science Stories: Teachers and Children as Science Learners* (Boston, Houghton Mifflin), an elementary science methods textbook. Her research explores broadening the participation of girls and young women in science through understanding the complexities of meaning making in science as situated in contemporary culture.

Nasir Mahmood is Lecturer in Science Education at the Institute of Education and research, University of the Punjab, Lahore, Pakistan. He has taught science at the secondary level for six years.

Paolo Mascheretti is Professor of Physics in the Geology Department and in the graduate School for Physics and Mathematics Teaching of the University of Pavia, Italy. His research focuses on physics education in high school and on teacher preparation.

Elizabeth McKinley is Lecturer in Science Education at the Centre for Science, Mathematics and Technology Education Research at the University of Waikato, Hamilton, Aotearoa, New Zealand. Her research interests center around the influence of culture and language in science and science education. She is also interested in feminist issues as they relate to science education curriculum, policy, and development.

Sharon E. Nichols is Associate Professor of Science Education at East Carolina University, USA where she currently serves as graduate coordinator. Her research interests include school/university partnerships, case-based and narrative pedagogies, reflective practice, and international science education. She is currently involved in an international research collaboration with science educators from West Visayas State University in the Philippines.

James O'Callahan is Professor of Science at the Universidad de Andes in Merida, Venezuela.

Lesley Parker is a science educator and Senior Deputy Vice-Chancellor of Curtin University of Technology in Perth, Western Australia. She has served as Associate Director of the National Key Centre for School Science and Mathematics where she has been a leader in forging equity-related policies and research in science education.

Jon E. Pedersen is Professor of Science Education in the Department of Instructional Leadership and Academic Curriculum at the University of Oklahoma, USA where he teaches preservice and practicing K-12 teachers and graduate students in Science Education. His research focuses on Science/Technology/Society issues and episodic memory in science teacher preparation.

Marilu Rioseco iş a former science faculty member at the Universidad de Concepcion, Chile. She currently resides in Santiago, Chile where she continues her work with Chilean teachers in the area of physics and chemistry education.

Deborah J. Tippins is Associate Professor of Science and Elementary Education at the University of Georgia, USA. Her research interests include sociocultural dimensions of science teaching and learning, case-based and narrative pedagogies, reflective practice, alternative assessment, and international science education. She recently co-edited (with Tom Koballa) *Cases in Middle and Secondary Science Education: The Promises and Dilemmas*. She is involved in an international research collaboration with science teacher educators from West Visayas State University in the Philippines.

James J. Watters teaches science education methods courses to elementary preservice teachers at the Queensland University of Technology, Brisbane, Australia. He also teaches early childhood teachers in professional development courses. His research interests include science teacher professional development and learning in children, especially gifted children. In 1999 he received an award for his contributions to science education from the Science Teachers Association of Queensland.

Yehudith Weinberger is an instructor in the Kibbutzim College of Education, Israel, where she teaches preservice and practicing junior high school teachers in Science Education. Her main research interests are the development of learners' thinking and the education of science teachers--from their preservice preparation through their inservice advanced study.

Anat Zohar is Senior Lecturer in the School of Education at the Hebrew University, Jerusalem, Israel. Her main research interests are the development of students' reasoning skill, teachers' cognition in relation to instruction of higher order thinking, and gender issues in science education. She is the director of the "Thinking in Science Classrooms" project.

FIRST AUTHOR INDEX

SUBJECT INDEX

Science & Technology Education Library

Series editor: Ken Tobin, *University of Pennsylvania, Philadelphia, USA*

Publications

1. W.-M. Roth: *Authentic School Science.* Knowing and Learning in Open-Inquiry Science Laboratories. 1995 ISBN 0-7923-3088-9; Pb: 0-7923-3307-1
2. L.H. Parker, L.J. Rennie and B.J. Fraser (eds.): *Gender, Science and Mathematics.* Shortening the Shadow. 1996 ISBN 0-7923-3535-X; Pb: 0-7923-3582-1
3. W.-M. Roth: *Designing Communities.* 1997 ISBN 0-7923-4703-X; Pb: 0-7923-4704-8
4. W.W. Cobern (ed.): *Socio-Cultural Perspectives on Science Education.* An International Dialogue. 1998 ISBN 0-7923-4987-3; Pb: 0-7923-4988-1
5. W.F. McComas (ed.): *The Nature of Science in Science Education.* Rationales and Strategies. 1998 ISBN 0-7923-5080-4
6. J. Gess-Newsome and N.C. Lederman (eds.): *Examining Pedagogical Content Knowledge.* The Construct and its Implications for Science Education. 1999 ISBN 0-7923-5903-8
7. J. Wallace and W. Louden: *Teacher's Learning.* Stories of Science Education. 2000 ISBN 0-7923-6259-4; Pb: 0-7923-6260-8
8. D. Shorrocks-Taylor and E.W. Jenkins (eds.): *Learning from Others.* International Comparisons in Education. 2000 ISBN 0-7923-6343-4
9. W.W. Cobern: *Everyday Thoughts about Nature.* A Worldview Investigation of Important Concepts Students Use to Make Sense of Nature with Specific Attention to Science. 2000 ISBN 0-7923-6344-2; Pb: 0-7923-6345-0
10. S.K. Abell (ed.): *Science Teacher Education.* An International Perspective. 2000 ISBN 0-7923-6455-4

KLUWER ACADEMIC PUBLISHERS – DORDRECHT / BOSTON / LONDON